环境科学专业英语
Professional English for Environmental Science

主　编：杨金燕
副主编：何文艳　李金鑫

科学出版社
北　京

内 容 简 介

本书围绕环境科学领域的相关内容组织材料,共分为12个单元,内容包括能源、气候变化、土壤形成与侵蚀、大气污染、水污染等。每单元分5个部分,包括精读、泛读、对话、补充词汇及补充信息。精读部分为该领域的基础知识和侧重词汇;泛读部分为该领域的前沿或热点问题;对话部分旨在提高学生的英文语言表达能力,提供交流素材;补充词汇部分提供相关领域应掌握的基本词汇信息;补充信息部分包括专业词汇构词法、专业文献阅读方法、学术报告准备、专业演讲技巧、英文学术论文的写作方法等。

本书可作为高等院校环境科学与工程专业及相关专业的本科生及研究生的专业英语教材,也可供环境学领域相关的科研人员和相关工程技术人员参考使用。

图书在版编目(CIP)数据

环境科学专业英语 = Professional English for Environmental Science / 杨金燕主编. — 北京：科学出版社，2018.1（2024.7重印）
ISBN 978-7-03-055782-7

Ⅰ.①环… Ⅱ.①杨… Ⅲ.①环境科学–英语–教材 Ⅳ.①X

中国版本图书馆 CIP 数据核字（2017）第 298741 号

责任编辑：张　展　刘莉莉 / 责任校对：彭　映
责任印制：罗　科 / 封面设计：墨创文化

科学出版社出版
北京东黄城根北街16号
邮政编码：100717
http://www.sciencep.com

成都锦瑞印刷有限责任公司 印刷
科学出版社发行　各地新华书店经销

*

2018年1月第 一 版　开本：B5（720×1000）
2024年7月第五次印刷　印张：17
字数：340 000
定价：49.00元
（如有印装质量问题，我社负责调换）

环境科学专业英语

主　　编：杨金燕

副 主 编：何文艳　李金鑫

参编人员：廖瑜亮　朱燕园　王　梅　于雅琪
　　　　　罗后巧　罗　玲　苟　敏　庞丽娜

前　　言

环境是人类生存和经济发展的基本前提。而随着社会经济的快速发展，环境问题已成为人类生存和发展所不可回避的重要问题。保护环境，遏制生态恶化，已成为社会和政府共同面临的重要任务。因此，需要环境类专业学者和广大环保工作者深入了解国内外环境领域研究现状，明确国际环境领域热点问题，掌握前沿环境保护技术。

随着学术研究的国际化和国际交流规模的扩大，提高国际学术交流能力十分必要。因此，专业英语阅读及表达能力的提升是大势所趋。此外，随着我国环保事业的发展，需要培养的环保技术人才越来越多。因而，需要一本既能介绍环境科学基础知识，又能反映当前热点问题的专业英语教材，以满足新时期环境专业人才培养的教学需求，并为环保工作者提供参考。

本教材围绕环境科学领域的相关内容组织材料。内容基本涵盖了环境科学专业各基础领域，包括能源、气候变化、土壤形成与侵蚀、大气污染、水污染、有机污染、无机污染、固体废弃物污染、生态学、环境生物学、环境健康、环境监测、环境法律法规等各个领域。通过对范文的理解、词汇的记忆、长难句的分析和学术问题的讨论，读者可以了解环境科学专业的基础知识和前沿领域。通过对本教材的学习，可以掌握环境科学专业的英语表达方法，掌握一定的科技文献阅读、翻译、写作和学术演讲技巧，提高专业素质、工作技能和国际交流能力。

本教材是在四川大学环境科学专业本科生和研究生专业英语课程讲稿的基础上做了大量修改后完成的。全书共有 12 个单元，各单元的主体部分分别从环境专业的基础和前沿着手，进行多角度阐述，对环境专业本科阶段可能涉及的环境领域知识进行介绍，并辅以相关阅读材料，补充专业词汇，同时对科技英语的阅读和写作进行介绍。每个单元分为 5 个部分，分别为精读、泛读、对话、补充词汇及补充信息。精读部分为该领域的基础知识，侧重词汇和翻译，适合环境专业本科生学习；泛读部分为该领域的前沿或热点问题，适合环境专业研究生学习；对话部分旨在提高学生的英文语言表达能力，提供交流素材和表达方法；补充词汇部分提供相关领域应掌握的基本词汇信息；补充信息部分包括专业词汇构词法、专业文献阅读方法、学术报告准备、专业演讲技巧、英文学术论文的写作方法等内容。各单元后均有课后练习供读者使用。

本书由杨金燕主编，王梅、于雅琪、苟敏、何文艳、李金鑫、罗后巧、廖瑜亮和庞丽娜协助完成书中 Part B、Part C 和 Part D 部分材料的收集工作。感谢李金鑫、何文艳、朱燕园、廖瑜亮、罗玲对本书编写提供的帮助，尤其感谢何文艳、李金鑫付出了大量时间进行校对工作。在本书的编写过程中，参考了一些相关文献，在此向其著作者表示感谢！

本书可作为高等院校环境科学与工程专业及相关专业的本科生及研究生的专业英语教材，也可供环境学领域相关科研人员和工程技术人员参考使用。

受编者知识水平及教材篇幅所限，书中难免存在疏漏和不当之处，敬请各位专家、学者和广大读者斧正，以便编者在今后的教学和科研工作中不断改进和提高。

编　者
2017 年 10 月

Contents

Unit 1 Basics of Environmental Science ·········· 1
   Part A Intensive Reading   Multidisciplinary Nature of Environmental Studies ···· 1
   Part B Extensive Reading   Environmental Interactions, Cycles and Systems ······ 4
   Part C Dialogue ·········· 11
   Part D Supplementary Vocabulary ·········· 13
   Part E Supplementary Knowledge   科技英语词汇的特点、构成及其翻译 ···· 14

Unit 2 Environment and Ecology ·········· 18
   Part A Intensive Reading   Ecosystems ·········· 18
   Part B Extensive Reading   Ecology and Environmentalism ·········· 24
   Part C Dialogue ·········· 29
   Part D Supplementary Vocabulary ·········· 33
   Part E Supplementary Knowledge   无机化合物命名法 ·········· 34

Unit 3 Energy and Climate Change ·········· 38
   Part A Intensive Reading   Overview of Climate Change Science ·········· 38
   Part B Extensive Reading   Energy, Climate Change and Environment ·········· 42
   Part C Dialogue ·········· 47
   Part D Supplementary Vocabulary ·········· 50
   Part E Supplementary Knowledge   有机化合物命名法 ·········· 51

Unit 4 Air Pollution ·········· 57
   Part A Intensive Reading   Air Pollution ·········· 57
   Part B Extensive Reading   Industrial Air Pollution Leaves Magnetic Waste in
                                   the Brain ·········· 63
   Part C Dialogue ·········· 67
   Part D Supplementary Vocabulary ·········· 69
   Part E Supplementary Knowledge   1. Air Pollution Prevention and Control
                                            Action Plan
                                            2. 如何筛选和阅读文献 ·········· 71

Unit 5 Water Pollution ·········· 78
   Part A Intensive Reading   Water Pollution ·········· 78

Part B	Extensive Reading	Micropollutant Fate in Wastewater Treatment: Redefining "Removal"	84
Part C	Dialogue		88
Part D	Supplementary Vocabulary		92
Part E	Supplementary Knowledge	1. The Action Plan for Prevention and Control of Water Pollution	
		2. 学术演讲的内容和结构	94

Unit 6 Soil Formation and Erosion ········ 99

Part A	Intensive Reading	Weathering	99
Part B	Extensive Reading	Soil Formation, Ageing, and Taxonomy	104
Part C	Dialogue		110
Part D	Supplementary Vocabulary		112
Part E	Supplementary Knowledge	1. Action Plan for Prevention and Control of Soil Pollution	
		2. 如何做学术海报	116

Unit 7 Solid Waste ········ 124

Part A	Intensive Reading	Solid Wastes	124
Part B	Extensive Reading	Waste Production must Peak this Century	131
Part C	Dialogue		135
Part D	Supplementary Vocabulary		139
Part E	Supplementary Knowledge	英文学术论文的题目与摘要	140

Unit 8 Organic Pollution ········ 146

Part A	Intensive Reading	Pesticides	146
Part B	Extensive Reading	How to Clean a Beach	151
Part C	Dialogue		157
Part D	Supplementary Vocabulary		160
Part E	Supplementary Knowledge	如何写英文学术论文的引言	162

Unit 9 Heavy Metal Pollution ········ 167

Part A	Intensive Reading	Heavy Metals	167
Part B	Extensive Reading	Phytoremediation of Heavy Metals: A Green Technology	171
Part C	Dialogue		176
Part D	Supplementary Vocabulary		178
Part E	Supplementary Knowledge	如何写英文学术论文的材料与方法	180

Unit 10	**Environmental Biology**	185
Part A	Intensive Reading Microbial Composition and Stoichiometry	185
Part B	Extensive Reading Soil and Groundwater Treatment-Phytoremediation	189
Part C	Dialogue	199
Part D	Supplementary Vocabulary	201
Part E	Supplementary Knowledge 如何写英文学术论文的结果	202
Unit 11	**Environmental Health**	209
Part A	Intensive Reading Environmental Health	209
Part B	Extensive Reading Transforming Environmental Health Protection	213
Part C	Dialogue	220
Part D	Supplementary Vocabulary	223
Part E	Supplementary Knowledge 如何写英文学术论文的讨论	224
Unit 12	**Other Environmental Area**	229
Part A	Environmental Monitoring	229
Part B	Risk Assessment	235
Part C	Remote Sensing	240
Part D	Introduction to the Environmental Impact Assessment Process	246
Part E	Supplementary Knowledge 1. 英文学术论文中的参考文献和致谢 2. 实验室常用英语	251

Unit 1   Basics of Environmental Science

Part A   Intensive Reading

Multidisciplinary Nature of Environmental Studies

The word environment is derived from the French verb "environner" which means to "encircle or surround". Thus our environment can be defined as the physical, chemical and biological world that surrounds us as well as the complex of social and cultural conditions affecting an individual or community. This broad definition includes the natural world and the technological environment as well as the cultural and social contexts that shape human lives. It includes all factors living and nonliving that affect an individual organism or population at any point in the life cycle; set of circumstances surrounding a particular occurrence and all the things that surround us.

There are three reasons for studying the state of the environment. The first, is the need for information that clarifies modern environmental concepts like equitable use of natural resources, more sustainable life styles etc. Second, there is a need to change the way in which we view our own environment, using practical approach based on observation and self-learning. Third, there is a need to create a concern for our environment that will trigger environmental protection, including simple activities we can do in our daily life to protect it.

Environmental science is essentially the application of scientific methods and principles to the study of environmental issues, so it has probably been around in some forms as science itself. Environmental science is often confused with other fields of related interest, especially ecology, environmental studies, environmental education and environmental engineering. Environmental science is not constrained with any one discipline and it is a comprehensive field.

Environmental science is not ecology though that discipline may be included. Ecologists are interested in the interactions between some kind of organisms and their surroundings. Most ecological research and training do not focus on environmental problems except those problems impacting the organism of interest. Environmental scientists may or may not include organisms in their field of view. They mostly focus on the environmental problems which may be purely physical in nature.

Acid deposition can be studied as a problem of emissions and characteristic of the atmosphere without necessarily examining its impact on organisms.

There are two types of environment: natural environment and man-made environment.

Natural environment: The environment in its original form without the interference of human beings is known as natural environment. It operates through self-regulating mechanism called homeostasis i.e, and any change in the natural ecosystem brought about by natural processes is counter balanced by changes in other components of environment.

Man-made or anthropogenic environment: The environment changed or modified by the interference of human beings is called man-made environment. Man is the most evolved creature on this earth. He is modifying the environment according to his requirements without bothering for its consequences. Increasing technologies and population explosion are deteriorating the environment more and more.

Scope of Environmental Studies

Because, the environment is complex and actually made up of many different environments, including natural, constructed and cultural environments. Environmental studies are the inter disciplinary examination of how biology, geology, politics policy studies, law, religion engineering, chemistry and economics combine to inform the consideration of humanity's effects on the natural world. By studying environmental science, students may develop a breadth of the interdisciplinary and methodological knowledge in the environmental fields that enables them to facilitate the definition and solution of environmental problems.

The scope of environmental studies is that, the current trend of environmental degradation can be reversed if people of educated communities are organized and empowered; experts are involved in sustainable development. Environmental factors greatly influence every organism and their activities. The major areas in which the role of environmental scientists is of vital importance are natural resources, ecosystems, biodiversity and its conservation, environmental pollution, social issues and environment, human population and environment.

Environmental science is essentially a multidisciplinary approach and its components include Biology, Geology, Chemistry, Physics, Engineering, Sociology, Health Sciences, Anthropology, Economics, Statistics and Philosophy. An understanding of the working of the environment requires the knowledge from wide ranging fields.

Importance of Environmental Education

Environment is not a single subject. It is an integration of several subjects that include both science and social studies. To understand all the different aspects of our environment, we need to understand biology, chemistry, physics, geography, resources management, economics, and population issues. Thus, the scope of environmental studies is extremely wide and covers some aspects of nearly every major discipline.

We live in a world where natural resources are limited. Water, air, soil, mineral, oil, the products we get from forests, grasslands, oceans and from agriculture and livestock, are all parts of our life support systems. Without them, life itself would be impossible. If we use them more and more, the earth's resources must inevitably shrink. The earth cannot be expected to sustain indefinitely due to over utilization of resources, misuse of resources. We waste or pollute large amount of clean water. We discard plastic, solid wastes and liquid wastes from industries which cannot be managed by natural processes. These accumulate in our environment, leading to a variety of diseases and other adverse environmental impacts, now seriously affecting all our lives. Air pollution leads to respiratory diseases, water pollution leads to gastro-intestinal diseases and many pollutants are known to cause cancer.

This situation will only improve if each of us begins to take action in our daily lives that will help to preserve our environmental resources. We cannot expect government alone to manage the safeguarding of the environment, nor can we expect other people to prevent environmental damage. We need to do it ourselves. It is a responsibility that each of us must take on as one's own.

Environmental Issues of Global Concern

The main environmental issues today that are wide ranging and all-encompassing are deforestation, biodiversity loss, soil erosion, climate change, pesticide build-up, industrial and municipal pollution. All these problems can be categorized into three main issues:

1. Population explosion
2. Land degradation
3. Environmental pollution: industrialization, agriculture/fertilizer/pesticide/ greenhouse gases, air pollution, acid rain, ozone depletion, greenhouse effect, water pollution and deforestation.

Source: Prasadini P, Lakshmi G S. Environmental Science-Birm 301 Study Material.

Words and Phrases

ecology [ɪ'kɒlədʒɪ]　*n.* 生态学，生态
homeostasis [ˌhəʊmɪə'steɪsɪs]　*n.* 体内稳态，内环境稳定
evolve [ɪ'vɒlv]　*vt.* 发展，逐渐演变；使逐步形成；推断出；*vi.* 进化；进化形成
geology [dʒɪ'ɒlədʒɪ]　*n.* 地质学，地质
biodiversity [ˌbaɪə(ʊ)daɪ'vɜːsətɪ]　*n.* 生物多样性
livestock ['laɪvstɒk]　*n.* 牲畜，家畜
soil erosion [ɪ'rəʊʒn]　土壤侵蚀
acid deposition　酸沉降
original form　原始形式
population explosion　人口爆炸
multidisciplinary approach　多学科的方法
solid wastes　固体废物
respiratory diseases　呼吸道疾病
gastro-intestinal diseases　胃肠道疾病
safeguarding of the environment　环境保护
environmental damage　环境损害
climate change　气候变化
pesticide build-up　农药积聚
land degradation　土地退化

Questions

1. What is environment?
2. What are the environmental issues of global concern?
3. Please briefly describe the scope of environmental studies.
4. From your point of view, what are the environmental problems in your hometown?

Sentence-making

1. climate change, population explosion, environmental degradation
2. greenhouse effect, carbon dioxide, methane
3. acid rain, deposition, sulfur
4. alternative fuels, energy, pollution
5. deforestation, land degradation, soil erosion

Part B　Extensive Reading

Environmental Interactions, Cycles and Systems

Inquisitive children sometimes ask whether the air they breathe was once breathed by a dinosaur. It may have been. The oxygen that provides the energy to power your body has been used many times by many different organisms, and the carbon, hydrogen, and other elements from which your body is made have passed through many other bodies during almost four billion years when life has existed on our planet. All the materials found at the surface of the earth, from the deepest ocean trenches to the top of the atmosphere are engaged in cycles that move them from place to place. Even the solid rock beneath your feet moves, as mountains eroded, sedimentary rocks are subducted into the earth's mantle, and volcanic activity releases new igneous rock. There is nothing new or original in the idea of recycling!

The cycles proceed at widely differing rates that vary from one part of the cycle to another. Cycling rates are usually measured as the time a molecule or particle remains in a particular part of the cycle. This is called its "residence time" or "removal time". On average, a dust or smoke particle in the lower atmosphere (the troposphere) remains airborne for a matter of a few weeks at most before rain washes it to the surface, and a water molecule remains in the air for around 9 or 10 days. Material reaching the upper atmosphere (the stratosphere) resides there for much longer, sometimes for several years, and water that drains from the surface into ground water may remain there for up to 400 years, depending on the location.

Water that sinks to the bottom of the deep oceans eventually returns to the surface, but this takes very much longer than the removal of water molecules from the air. In the Pacific Ocean, for example, it takes 1000 to 1600 years for deep water to return to the surface, and in the Atlantic and Indian Oceans it takes around 500 to 800 years. This is relevant to concern about the consequences of disposing industrial and low-level radioactive waste by sealing it in containers and dumping them in the deep oceans.

Those monitoring the movement of materials through the environment often make use of label, different labels being appropriate for different circumstances. In water, chemically inert dyes are often used. Certain chemicals will bond to particular substances. When samples are recovered, analysis reveals the presence or absence of the chemical label. Radioisotopes are also used. These consist of atoms chemically identical to all other atoms of the same element, but with a different mass, because of a difference in the number of neutrons in the atomic nucleus. Neutrons carry no charge and so take no part in chemical reactions, the chemical characteristics of an element being determined by the number of protons, with a positive charge, in its atomic nucleus.

You can work out the atmospheric residence time of solid particles by releasing particles labelled chemically or with radioisotopes and counting the time that it takes for them to be washed back to the ground, although the resulting values are very approximately. Factory smoke belching forth on a rainy day may reach the ground within an hour or even less; the exhaust gases from an aircraft flying at high altitude will take much longer, because they are further from the ground to start with and in much drier air. It is worth remarking, however, that most of the gases and particles which pollute the air and can be harmful to health have very short atmospheric residence time. Sulphur dioxide, for example, which is corrosive and contributes to acid rain, is unlikely to remain in the air for longer than one month and may be washed to the surface within one minute of being released. The atmospheric residence time for water molecules is calculated from the rate at which surface water evaporates and returns as precipitation.

The deep oceans are much less accessible than the atmosphere, but water carries a natural label in the form of carbon-14 (^{14}C). This forms in the atmosphere through the bombardment of nitrogen (^{14}N) by cosmic radiation, but it is unstable and decays to the commoner ^{12}C at a steady rate. While water is exposed to the air, both ^{12}C and ^{14}C dissolve into it, but once isolated from the air the decay of ^{14}C means that the ratio of the two changes, ^{12}C increasing at the expense of ^{14}C. It is assumed that ^{14}C forms in the air at a constant rate, so the ratio of ^{12}C to ^{14}C is always the same and certain assumptions are made about the rate at which atmospheric carbon dioxide dissolves into sea water and the rate at which water rising from the depths mixes with surface water. Whether or not the initial assumptions are true, the older water is the less ^{14}C it will contain, and if the assumptions are true the age of the water can be calculated from its ^{14}C content in much the same way as organic materials are ^{14}C-dated.

Carbon, oxygen, and sulphur are among the elements living organisms use and they are being cycled constantly through air, water, and living cells. The other elements required as nutrients are also engaged in similar biogeochemical cycles. Taken together, all these cycles can be regarded as components of a very complex system functioning on a global scale. Used in this sense, the concept of a "system" is derived from information theory and describes a set of components which interact to form a coherent, and often self-regulating, whole. Your body can be considered as a system in which each organ performs a particular function and the operation of all the organs is coordinated so that you exist as an individual who is more than the sum of the organs from which your body is made.

The idea that biogeochemical cycles are components of an overall system raises an obvious question: what drives this system? It used to be thought that the global system is purely mechanical, driven by physical forces, and, indeed, this is the way it can seem. Volcanoes, from which atmospheric gases and igneous rocks erupt, are purely physical phenomena. The movement of crustal plates, weathering of rocks, condensation of water vapour in cooling air to form clouds leading to precipitation—all these can be explained in purely physical terms and they carry with them the substances needed to sustain life. Organisms simply grab what they need as it passes, modifying their requirements and strategies for satisfying them as best they can when conditions change.

Yet this picture is not entirely satisfactory. Consider, for example, the way limestone and chalk rocks form. Carbon dioxide dissolves into raindrops, so rain is very weakly acid. As the rain water washes across rocks it reacts with calcium and silicon in them to form silicic acid and calcium bicarbonate, as separate calcium and bicarbonate ions. These are carried to the sea, where they react to form calcium carbonate, which is insoluble and slowly settles to the sea bed as a sediment that, in time, may be compressed until it becomes the carbonate rock we call limestone. It is an entirely inanimate process. Or is it? If you examine limestone closely you will see it contains vast numbers of shells, many of them minute and, of course, often crushed and deformed. These are of biological origin. Marine organisms "capture" dissolved calcium and bicarbonate to "manufacture" shells of calcium carbonate. When they die the soft parts of their bodies decompose, but their insoluble shells sink to the sea bed. This appears to be the principal mechanism in the formation of carbonate rocks and it has occurred on a truly vast scale, for limestones and chalks are among the commonest of all sedimentary rocks. The famous White Cliffs of Dover are made from the shells of once-living marine organisms, now crushed, most of them beyond individual recognition.

Here, then, is one major cycle in which the biological phase is of such importance that we may well conclude that the cycle is biologically driven, and its role extends further than the production of rock. The conversion of soluble bicarbonate into insoluble calcium carbonate removes carbon, as carbon dioxide, from the atmosphere and isolates it. Eventually crustal movements may return the rock to the surface, from where weathering returns it to the sea, but its carbon is in a chemically stable form. Other sedimentary rock on the ocean floor is subducted into the mantle. From there its carbon is returned to the air, being released volcanically, but the cycle must be measured in many millions of years. For all practical purposes, most of the carbon is stored fairly permanently. As the newspapers constantly remind us, carbon dioxide is a "greenhouse gas", one of a number of gases present in the atmosphere that are transparent to incoming, short-wave solar radiation, but partially opaque to long-wave radiation emitted from the earth's surface when the sun has warmed it. These gases trap outgoing heat and so maintain a temperature at the surface markedly higher than it would be were they absent. Since the Earth formed, some 4.6 billion years ago, the Sun has grown hotter by an estimated 25 to 30 percent, and the removal of carbon dioxide from the air, to a significant extent as a result of biological activity, has helped prevent surface temperatures rising to intolerable levels.

Authorities differ in the importance they allot to the role of the biota (the total of all living organisms in the world or some defined part of it) in driving the biogeochemical cycles, but all agree that it is great, and it is self-evident that the constituents of the biota shape their environment to a considerable extent. Grasslands are maintained by grazing herbivores, which destroy seedlings by eating or trampling them, so preventing the establishment of trees, and over-grazing can reduce semi-arid land to desert. The presence of gaseous oxygen in the atmosphere is believed to result from photosynthesis.

We alter the environment by the mere fact of our existence. By eating, excreting, and breathing we interact chemically with our surroundings and thereby change them. We take and use materials, moving them from place to place and altering their form. Thus we subtly modify environmental conditions in ways that favour some species above others. In our concern that our environmental modifications are now proceeding on such a scale as to be unduly harmful to other species and possibly ourselves, we should not forget that in this respect we differ from other species only in degree. All living things alter their surroundings, through their participation in the cycles that together comprise the system which is the dynamic earth.

Source: Allaby M. Basics of Environmental Science, 2nd Edition[M].The Taylor & Francis e-Library, 2002.

Words and Phrases

hydrogen ['haɪdrədʒən]　　*n.* [化学] 氢；氢气
[例句] Water can be reduced to oxygen and hydrogen by electrolysis.
erode [ɪ'rəʊd]　　*v.* 侵蚀；损坏
[例句] I believe that anywhere within a metre or so of daily maximum sea level would be swiftly eroded.
mantle ['mæntl]　　*n.* 地幔；斗篷；覆盖物
[例句] It is this shoving of the mantle that created the volcanoes of the Mediterranean.
troposphere ['trɒpəsfɪə(r); 'trəʊp-]　　*n.* [气象] 对流层
[例句] The lowest layer of the atmosphere is known as the troposphere, which is constantly in motion, causing the weather.

stratosphere ['strætəsfɪə(r)]　　*n.* 同温层；平流层
[例句] The amount of water in the stratosphere could change, and so the link between the ozone problem and the climate problem becomes important and could accelerate or delay the recovery of ozone.

drain [dreɪn]　　*n.* 排水沟；消耗；*v.* 放出，排出
[例句] All of this information will help scientists understand why the Pine Island Glacier drains so much ice to the ocean and how much it could contribute to sea level rise in the future.

radioisotope [ˌreɪdɪəʊˈaɪsətəʊp]　　*n.* [核] 放射性同位素
[例句] Cancer risk is determined by the age at exposure and concentration of radioisotopes in particular tissues.

atom ['ætəm]　　*n.* [物理] 原子
[例句] All substances, whether they are gaseous, liquid or solid, are made of atoms.

corrosive [kəˈrəʊsɪv]　　*adj.* 腐蚀的；侵蚀性的；*n.* 腐蚀物
[例句] The corrosive effects of water in those conditions would outpace Bermudan reefs' ability to grow.

evaporate [ɪˈvæpəreɪt]　　*v.* 蒸发；消失；衰退
[例句] Water in the sand between the two pots evaporates through the surface of the larger pot, where drier outside air is moving.

sulphur ['sʌlfə(r)]　　*n.* 硫；硫磺；硫磺色
[例句] Concentrations of smoke and major air pollutants such as sulphur dioxide have now decreased, mainly due to less coal burning by industry and homes.

organism ['ɔːgənɪzəm]　　*n.* 有机体；生物体；微生物
[例句] All living organisms have to adapt to changes in environmental conditions.

bombardment [bɒmˈbɑːdm(ə)nt]　　*n.* 轰炸；炮击
[例句] The town lay in ruins after a heavy bombardment.

self-regulating ['self 'regjuleɪtɪŋ]　　*adj.* 自制的；自动调节的
[例句] This activates the self-regulating systems in our living tissue so as to alleviate the bodily imbalance and lead to recovery.

vapour ['veɪpə(r)]　　*n.* 蒸气（等于 vapor）；水蒸气
[例句] These winds can transport large amounts of water vapour over hundreds of kilometres.

limestone ['laɪmstəʊn]　　*n.* [岩] 石灰岩
[例句] In return for a cozy, safe place to live, the algae provide the building blocks that polyps need to survive and make limestone to build the reef structures.

calcium ['kælsɪəm]   *n.* [化学] 钙
[例句] What you eat and drink, from childhood on, is critical to the amount of calcium in your bones.

silicon ['sɪlɪkən]   *n.* [化学] 硅；硅元素
[例句] But the major ingredient of semiconductors is silicon, the second most abundant element on earth.

sediment ['sedɪmənt]   *n.* 沉积；沉淀物
[例句] Ocean plants buried in sediment can help reveal Earth's temperature thousands of years ago.

inanimate [ɪn'ænɪmət]   *adj.* 无生命的；无生气的
[例句] So we've got a possible explanation of the difference between an animated and an unanimated or an inanimate body to it.

bicarbonate [ˌbaɪ'kɑ:bənət]   *n.* 碳酸氢盐；重碳酸盐；酸式碳酸盐
[例句] The doctors used magnetic resonance imaging to see how much of the tagged biocarbonate was converted into carbon dioxide within the tumor.

decompose [ˌdi:kəm'pəʊz]   *v.* 分解；使腐烂
[例句] Most of the carbon is returned to near-surface waters when phytoplankton are eaten or decomposed, but some falls into the ocean depths.

biota [baɪ'əʊtə]   *n.* 生物区（系）；一时代（一地区）的动植物
[例句] Soil microbe is a key functional factor in soil biota, and some microbiological parameters can be cited as general biological indicators for assessing soil health.

constituent [kən'stɪtjuənt]   *n.* 成分
[例句] In initial tests, they will focus on breaking down the polymer into its constituents.

herbivore ['hɜ:bɪvɔ:(r)]   *n.* [动] 食草动物，植食动物；食草者
[例句] These beetles subsist entirely on the undigested nutrients in the waste of herbivores like sheep, cattle, and elephants.

photosynthesis [ˌfəʊtəʊ'sɪnθəsɪs]   *n.* 光合作用
[例句] This action all takes place inside tiny capsules called chloroplasts that reside inside every plant cell — and which is where photosynthesis happens.

dynamic [daɪ'næmɪk]   *adj.* 动态的；动力的；动力学的；*n.* 动态；动力
[例句] The routing criteria can be static or dynamic.

ocean trench	海沟	**volcanic activity**	火山活动
sedimentary rock	沉积岩	**igneous rock**	火成岩
Earth's mantle	地幔	**residence time**	停留时间；滞留时间

removal time　拆卸时间
Pacific Ocean　太平洋
Atlantic Ocean　大西洋
Indian Ocean　印度洋
radioactive waste　放射性废弃物
chemically inert dye　惰性染料
atomic nucleus　原子核
sulphur dioxide　二氧化硫
cosmic radiation　宇宙辐射
carbon dioxide　二氧化碳
crustal plate　地壳板块
chalk rock　白垩岩
silicic acid　硅酸
calcium bicarbonate　碳酸氢钙
bicarbonate ion　碳酸氢根离子
solar radiation　太阳辐射

Questions

1. What determines the rate of circulation?
2. What are the factors that affect the residence time?
3. Which method is usually used to calculate the water and atmospheric residence time? Why use these labels to monitor the environment?
4. Talk about your understanding about the environmental system.
5. What is the driving force of the global system?
6. Describe the process of the water cycle and the carbon cycle in your own words.

Part C Dialogue

Dialogue 1

A: I heard that you recently read some books about environmental science, can I ask you some questions?

B: Of course.

A: What are the environmental problems in our life? Can you explain to me in several different aspects?

B: Environmental problems can be divided into four categories: primary environment problems, secondary environment problems, pollution of the environment problems and environmental interference problems. Native environment is naturally produced by the initial environmental problems.

A: Like volcanoes, earthquakes, typhoons, etc?

B: You're right. Secondary environmental problems follow the primary environmental problems, such as soil erosion, desertification, salt marshes, species extinction, etc. The last two environmental problems relate to human activities, such as water, atmosphere, soil pollution, and noise, vibration, electromagnetic interference, thermal interference, etc.

A: How the environment problems get the attention of the people?

B: In 1962, the United States, a Marine biologist Rachel Carson published the book *Silent Spring*, and it popularly shows the serious ecological damage of pesticide use, marking the beginning of modern environmental science.

A: What are the most serious environmental problems today?

B: Currently, global environmental problems that the international community care most about are: greenhouse effect, water crisis, biodiversity decrease, forest decline, acid rain pollution, ozone depletion, land desertification and garbage plague, etc.

A: For instance the greenhouse effect, how does the global environment problem form?

B: On the one hand, human activities are strongly changing the composition of earth's atmosphere, emitting more and more gas. Most gas can absorb infrared radiation and thereby warm itself. By this way the earth's average temperature will gradually rise, among various greenhouse gas the most notable is the role of carbon dioxide.

A: As far as I see, on the other hand, the law of the earth's overall environmental evolution has a close relationship with the Greenhouse Effect. As we all know, ice age appears several times in the long history of the earth.

B: Now you have got an overview of environmental science.

<small>Source: http://www.globalissues.org/issue/168/environmental-issues</small>

Dialogue 2

A: What did you do over the weekend?

B: I went a global warming rally in London. It was fantastic to be around so many people who care about the environment.

A: Do you think there's anything we can do to reverse the damage that's been done already?

B: It might not be possible to fix the problems that we've created for ourselves, but there are lots of things we can do to prevent more damages from happening.

A: Like what?

B: Well, we can use public transport instead of taking our cars for a start.

A: What else can we do to protect the environment?

B: If you do have to drive, you should make sure that your car runs on unleaded petrol. Also, your home should use sources of renewable energy.

A: How about recycling? Does that actually help?

B: Yes. You should take your glass, paper, plastic, cardboard, and tin cans to a recycling center.

A: What do you think is the biggest worry for our future?

B: I think that the issue of greatest concern is having enough sources of clean water for everyone.

A: I care more about our energy, and as the global population grow so fast, energy lack will surely be the primary problem.

B: Energy problem requires attention from the human society. But clean water is much more emergent and irreversible.

A: I had no idea you were such an environmentalist before!

B: To be honest, in order for the earth to continue to be a habitable place, we're all going to have to become more interested in the environment.

Source: http://talk.kekenet.com/show_1409

Part D　Supplementary Vocabulary

环境科学专业相关课程名称

Air Pollution Control Engineering 大气污染防治工程
Analytical Chemistry 分析化学
Botany 植物学
Cleaner Production 清洁生产
Dendrology 树木学
Environics 环境学
Ecological Restoration Engineering 生态恢复工程学
Environmental Acoustics 环境声学
Environmental Aerodynamics 环境空气动力学
Environmental Analytical Chemistry 环境分析化学
Environmental Biology 环境生物学
Environmental Bioremediation Technology 环境生物修复技术
Environmental Chemistry 环境化学
Environmental Data Analysis 环境数据分析
Environmental Ecology 环境生态学
Environmental Economics 环境经济学
Environmental Electromagnetics 环境电磁学
Environmental Ethics 环境伦理学
Environmental Engineering 环境工程学
Environmental Epidemiology 环境流行病学
Environmental Fluid Mechanics 环境流体力学
Environmental Geochemistry 环境地球化学
Environmental Geology 环境地学
Environmental Impact Assessment 环境影响评价

Environmental Information System 环境信息系统
Environmental Law 环境法学
Environmental Management Science 环境管理学
Environmental Materials 环境材料学
Environmental Medicine Detection 环境医学检测
Environmental Microbiology 环境微生物学
Environmental Monitoring 环境监测
Environmental Oceanography 环境海洋学
Environmental Optics 环境光学
Environmental Philosophy 环境哲学
Environmental Planning 环境规划学
Environmental Pollution Chemistry 环境污染化学
Environmental Restoration 环境修复技术
Environmental Simulation Science 环境模拟学
Environmental Soil Science 环境土壤学
Environmental Statistics 环境统计学
Environmental Systems Engineering 环境系统工程
Environmental Toxicology 环境毒理学
Hydraulics 水力学
Inorganic Chemistry 无机化学
Introduction to Environmental Science 环境学导论
Investigation of Pollution Source 污染源调查
Landscape Ecology 景观生态学
Measurement Science 测量学
Mine Environmental Economic 矿山环境经济
Mine Environmental Planning 矿区环境规划
Modern Environmental Science and Technology 现代环境科学技术
Noise Control 噪声控制
Organic Chemistry 有机化学
Physical Chemistry 物理化学
Pollution Meteorology 污染气象学
Soil and Water Conservation 水土保持学
Solid Waste Treatment and Application 固体废物的处理和应用
Water Pollution Control Engineering 水污染防治工程
Water Supply and Drainage Engineering 给排水工程

Part E Supplementary Knowledge

科技英语词汇的特点、构成及其翻译

一、科技英语词汇的特点

科技英语与普通英语在语音、语法等方面无显著不同，但科技英语是英语的一种特殊文体，在具体范围内，在一定的语言场合里，表达事物的定义、概念、含义、分类、假说、比较、因果及论证等时，所担负的修辞功能、出现频率、要求语言所表现的句子结构，都具有自身的特点。

科技英语词汇概括起来主要有 5 个特点：

(1) 多源于拉丁语和希腊语。

(2) 词义专一。

(3) 前、后缀出现频率高。

(4) 专业词汇出现的频率低。

(5) 广泛使用缩略词。

科技英语与不具有特殊色彩的核心英语语言基础是一致的。即，科技英语与普通英语在语音、语调、基本词汇和基本语法上没有不同，科技英语本身没有独特或特殊的语法现象，只是某些语言结构所担负的修辞功能和语言结构出现的频率不同。而科技英语词汇在特定语言环境下有其特殊的含义，它的构成也是有一定的规律可循。

二、科技英语词汇的分类

（一）功能词(functional words)

这类词汇均为英语的基本词汇，它适用于各种文体。包括介词、动词、连词、冠词、代词、副词等，如 we、must、the、without、will、start 等都是功能词。在普通英语文章与科技英语文章中，二者在意义上是一致的。

（二）次技术词(semi-technical words)

这类词汇也是科普英语中通用的词汇，各专业中出现频率很高，如 mechanical、science、technology、project 等。其中一部分借自普通词汇，其含义与原意有所不同。如 monitor 在科技英语中的词义是监视器。

另外有一些科技词汇是从普通英语词汇逐渐演变而来的，这类科技英语词汇中有一部分往往与普通英语词汇有着密切的关系，主要体现在物体外部形状相似、物体功能相似和物体之间的关系相似，这类词汇主要有以下三种：

(1) **物体外部形状相似**。这类词汇起初多用于英语口语中，以后根据约定俗成的原则，逐渐广泛地用于科技词汇中。如：wing 在科技英语中词义为机翼，leg 在科技英语中词义为支架。

(2) **物体的功能相似**。如：monkey 在科技英语中词义为打桩锤，pecker 为穿孔器。

(3) **物体之间的关系相似**。如：baby 普通词义为"婴儿"，而在科技英语中则表示"微小的物体"，如：babycar(微型汽车)，babycarrier(轻型航空母舰)，babytower(小型蒸馏塔)；mother 为"母亲"，在科技术语中表示"母体，主"，如：mother machine(工作母机)，mother rod[主(母)连杆]，mother-board(母板)；daughter 为"女儿"，在科技英语中表示"子体"，如：daughter board(子插件)，

daughter neutron（派生中子），daughter nucleus（子核）等。对于这类词汇，通过形象记忆会得到事半功倍的效果。

此外，科技英语中有些技术词汇的元音字母与普通英语某些单词中出现的元音字母有不同的用处。如：sleep（睡觉）—asleep（睡着的）；fire（点燃）—afire（燃烧着）。在这些普通词中的"a"表示原词的状态，但在科技英语中，字母"a"加"a"后的单词词义与原词词义相反。如：dynamic（动态的）—adynamic（非动态的）；static（静止的）—astatic（非静止的）。

（三）技术词汇（technical words）

在科技英语中，大部分技术词汇词义专一，一般都是一词一义，词形越长，词义越专一。如：nitroglycerin（硝酸甘油）等。

三、科技英语词汇的构成

技术词汇的构成主要有合成法、转化法、派生法、缩写法等，前两种和普通词汇构词法基本相同。派生法是在词根上添加前缀或后缀构成一个新词的方法。

科技文献中，由数词构成的合成词主要由希腊语和拉丁语的数量前缀来表示。这类词在化学词汇方面比较多。

科学技术的发展日新月异，缩写词也越来越多，科技英语词汇的缩写与普通英语词汇的缩写比较，有其独特的方式。有的缩写词是从两个单词中抽出部分字母而构成新词。如：

  reduction oxidation → redox（氧化还原）

  telegraph exchange → telex（电传）

  rectifier transformer → rectiformer（整流变压器）

有的缩写词是将词组中的每个词的首字母加在一起构成新词。如：

  light amplification by stimulated emission of radiation →laser（激光）

  sound navigation ranging → sonar（声呐）

  radio detecting and ranging → radar（雷达）

这类词汇的首字母必须以缩写词为单位构成新词，并按其字母的语言结构读音。如：laser[leiz]、radar['reɪdɑ:(r)]、sonar['səunɑ:]。

此外，还有一些缩写是将每个单词的第一个字母抽出构成缩写，但没有单词读音。只是读作字母，这一部分早已被大部分专业人员所接受，特别是在科技口语方面，人们为了避免使用那些又长又难上口的科技词汇，往往只说其缩写形式。既简单、易于上口，又能表达准确的含义。如：

  GC→gas chromatograph（气相色谱仪）

  PPM→parts per million（百万分率）

  GMT→Greenwich Mean Time（格林尼治平均时间）

  TDA→toluene diamine（甲苯二胺）

四、科技英语词汇的翻译

在科技英语的实际应用中，人们应特别注意在不同专业里具有不同含义的词汇。如："accumulator"在仪表专业中其含义为"累加器"，而在加工方面就是指"活套塔"；"reactor"在冶金方面指的是"反应炉"，在化工方面指的是"反应器或反应堆"，而在电气方面则指的是"电抗器"。所以，对于这类技术词汇，只有通过对上下文和具体语言环境的了解，才能确定其真正的含义，不能以此类推，否则就会产生误译。

在科技词汇中还有一些词汇看似非常简单，却很容易误译，如：bluegas（水煤气），常误译为"蓝色气体"；mild steel（低碳钢），有人误译为"暖钢"；uptake（直升烟道），有时误译为"举起"等。这类错误的出现主要犯了望文生义和不了解技术词汇的特有含义，碰到这类词汇，应勤于在专业技术词典里查找其真正含义或向同行请教，同时还应了解一些科技方面的有关专业知识，掌握大量的科技词汇，才能做到翻译准确。

综上所述，首先应该掌握功能词汇，它们是英语语言的基本词汇，适用于各种文体，是学习语言的基础；其次，应掌握好次技术词汇，它们是由普通词汇逐渐发展而来的，其中有一部分与原词含义有着密切的关系，是可以通过形象记忆或逻辑推理的；再加上对前缀、后缀的了解，往往可以举一反三，有一定的规律可循。

Reference: 艾蓬. 科技英语词汇的特点、构成及其翻译[J]. 兰州文理学院学报（社会科学版），2001(S4)：88-91.

Unit 2   Environment and Ecology

Part A   Intensive Reading

Ecosystems

An ecosystem is a community of living (biotic) organisms (plants, animals and microbes) in conjunction with the nonliving components (abiotic) of their environment (air, water and mineral soil), interacting as a system. An ecosystem is a biological community along with its physical environment, and comes in various sizes from limited spaces to the entire planet. Ecosystems are dynamic with networks of interactions among organisms, between organisms and their environment. They are linked together through nutrient cycle and energy flow. Ecosystems are controlled both by external factors (climate, the parent material which forms the soil and topography) and internal factors (decomposition, root competition or shading, disturbance, succession and types of species).

Ecosystems provide a variety of goods and services to human society. It is important for environmental engineers to recognize and understand a wide range of ecosystem services in environmental engineering activities (e.g., in order to meet the drinking water standard, it may be more cost-effective for a water utility to invest in natural capital to improve the ecosystems of water sources than to build an expensive new water filtration plant).

Energy Flow

Living organisms require two things from the environment: energy to provide power and nutrients to provide substance. Energy flow, also called the calorific flow, refers to the flow of energy through a food chain. The sun is responsible for virtually all of the earth's energy, which constantly gives the planet energy in the form of light while it is eventually used and lost in the form of heat throughout the trophic levels of a food web. The flow of energy in an ecosystem is an open system (i.e., it does not cycle and is converted to heat and lost for useful purposes forever).

The trophic (derived from the Greek referring to food) level of an organism is the position it occupies in a food chain. Based on the way to get food, organisms are classified as producers, consumers and decomposers. **Producers** (autotrophs) are typically plants or algae that do not usually eat other organisms, but pull nutrients from the soil or the ocean and manufacture their own food using photosynthesis powered by solar energy. An exception occurs in deep-sea hydrothermal ecosystems with no sunlight and chemosynthesis is used by organisms to make food. Since they are at the lowest trophic level, they are called primary producers. Higher up on the food chain, **consumers** (heterotrophs) are animals which cannot manufacture their own food and need to consume other organisms. Animals that eat primary producers (like plants) are called herbivores and animals that eat other animals are called carnivores, and animals that eat both plants and other animals are called omnivores. **Decomposers** (detritivores, such as bacteria and fungi) break down dead plants, animals and their wastes as energy and nutrients into the ecosystem for recycling.

There are five trophic levels. Decomposers are often left off food webs and trophic levels, but if included, they mark the end of a food chain.

Level 1: Plants and algae are primary producers.

Level 2: Herbivores are primary consumers.

Level 3: Carnivores are secondary consumers if they eat herbivores.

Level 4: Carnivores are tertiary consumers if they eat other carnivores.

Level 5: Apex predators are at the top of the food chain if they have no predators.

A general energy flow in ecosystems starts with fixation of solar energy by photoautotrophs (i.e., primary producers). Primary consumers (i.e., herbivores) absorb most of the stored energy in the plant through digestion, and transform it into the form of energy they need through respiration. The received energy is stored as body mass with some converted to body heat radiated away and some lost by the expulsion of undigested food via excretion or regurgitation. Secondary consumers (carnivores) then consume the primary consumers and absorb the energy embodied in the primary consumers through the process of digestion. As with primary consumers, some energy is lost from the system. There may be higher level consumers to repeat the processes further on. A final link in the food chain is decomposers which break down the organic matter of the dead consumers (at all levels) and the undigested food excreted by the consumers, and release nutrients into the environment.

Energy transfer between trophic levels is generally inefficient with about 90% energy lost for each transfer. Generally, primary consumers get about 10% of the energy produced by autotrophs, while secondary consumers get 1% and tertiary consumers get 0.1%. This means the top consumer of a food chain receives the least energy, as a lot of the food chain's energy has been lost to the environment instead of being absorbed between trophic levels. This loss of energy at each level limits typical food chains to only four to six links.

Nutrient Cycles

Nutrient cycle is the movement and exchange of organic and inorganic matter back into the production of living matter. It is a part of larger biogeochemical cycles involving macros nutrients (such as carbon, nitrogen, oxygen, phosphorus, and sulphur, etc.) and micros nutrients (iron, copper, sodium, etc.)—the important chemicals used as nutrients in ecosystems by living organisms. In contrast to the energy flow, the nutrient flow is a part of a closed system and these chemicals are recycled and replenished constantly instead of being lost. The chemicals are sometimes held for long periods of time in one place called a reservoir (e.g., coal deposits with carbon), or are held for only short periods of time in exchange pools (e.g., plants and animals). In this section, only some key macros nutrients are described because they play an essential role in ecosystems.

Oxygen Cycle

The oxygen cycle describes the movement of oxygen within its three main reservoirs: biosphere (the smallest of the three reservoirs with an average residence time of 50 years), atmosphere (it is 100 times of the oxygen mass in the biosphere with an average residence time of 4500 years), and lithosphere (it is 200 times of the oxygen mass in the atmosphere, the largest reservoir with an average residence time of 500 million years). Therefore, 99.5% of oxygen is stored in the lithosphere, 0.5% in the atmosphere and only 0.005% in the biosphere.

The gain of oxygen in the atmosphere is driven mainly by the sun through photosynthesis (55% by land plants and 45% by ocean phytoplankton). The losses of atmospheric oxygen are mainly by aerobic respiration at 77% (consumed by animals and bacteria), microbial oxidation 17%, fossil fuel combustion 4%, and photochemical oxidation 2%.

Carbon Cycle

The carbon cycle describes the movement of carbon within its main reservoirs: atmosphere, hydrosphere (oceans), lithosphere and biosphere. The lithosphere has the largest store of carbon (it collects its carbon from the atmosphere by the accumulated dead life form and releases its carbon by either slow geological movement or fast combustion as fuel by humans). Oceans are the 2nd largest reservoir for carbon (mainly in inorganic form). The biosphere on land has a much smaller carbon store (a small fraction of that in oceans) and the atmosphere has the smallest carbon store (about 1/3 of the biosphere). The ocean plays a vital role in the earth's carbon cycle, but it is the increasing carbon in the atmosphere that is of major concern in modern times due to the burning of fossil fuels.

In the past two centuries, human activities have seriously altered the global carbon cycle, most significantly in the atmosphere. Carbon in the earth's atmosphere exists in two main forms: carbon dioxide and methane. Both of these gases are partially responsible for the greenhouse effect. Methane produces a large greenhouse effect per volume as compared to carbon dioxide, but it exists in much lower concentrations and is more short-lived than carbon dioxide, making carbon dioxide the more important greenhouse gas of the two.

In the biosphere, carbon is mainly absorbed in the form of carbon dioxide by plants. Carbon is also released from the biosphere into the atmosphere in the course of biological processes. Aerobic respiration converts organic carbon into carbon dioxide and anaerobic respiration converts it into methane. After respiration, both carbon dioxide and methane are typically emitted into the atmosphere. Organic carbon is also released into the atmosphere via burning. While organic matter in animals generally decays quickly, releasing much of its carbon into the atmosphere through respiration, carbon stored as dead plant matter can stay in the biosphere for as much as a decade or more. Different plant types of plant matter decays at different rates (for example, woody substances retain their carbon longer than soft, leafy materials). Active carbon in soils can stay sequestered for up to a thousand years, while inert carbon in soils can stay sequestered for more than a millennium.

Nitrogen Cycle

The nitrogen cycle is the process by which nitrogen is converted between its various chemical forms (Figure1). Nitrogen is required to biosynthesis basic building blocks of plants, animals and other life forms (e.g., nucleotides for DNA and RNA and amino acids for proteins). Important processes in the nitrogen cycle include fixation, ammonification, nitrification, and denitrification.

Although 78% of the air is nitrogen, atmospheric nitrogen must be processed or "fixed" into ammonia (NH_3) to be used by plants. Some fixation occurs in lightning strikes, but most fixation is done by free-living or symbiotic bacteria. When a plant or animal dies, or an animal expels waste, the initial form of nitrogen is organic. Bacteria (or fungi in some cases) convert the organic nitrogen within the remains back into ammonium (NH_4^+), a process called ammonification or mineralisation. Nitrification is a two-step process with the biological oxidation of ammonia (with oxygen) into nitrite followed by the oxidation of these nitrites into nitrates (by two types of organisms that exist in most environment). Nitrification is important in agricultural systems, where fertilizer is often applied as ammonia. Denitrification is the reduction of nitrates back into the largely inert nitrogen gas (N_2), completing the nitrogen cycle. Nitrification (together denitrification) also plays an important role in the removal of nitrogen from municipal wastewater.

Human impact on the nitrogen cycle is diverse. Agricultural and industrial nitrogen inputs to the environment currently exceed inputs from natural nitrogen fixation and the global nitrogen cycle has been significantly altered over the past century. Human activities such as fossil fuel combustion, use of artificial nitrogen fertilizers, and release of nitrogen in wastewater have dramatically altered the global nitrogen cycle.

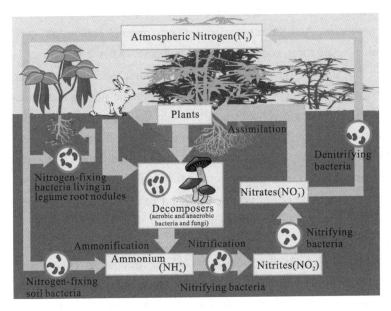

Figure 1 Nitrogen cycle (From: Wikipedia)

Phosphorus Cycle

The phosphorus cycle describes the movement of phosphorus through the lithosphere, hydrosphere, and biosphere. Unlike many other biogeochemical cycles, the atmosphere does not play a significant role in the movement of phosphorus, because phosphorus does not enter the atmosphere and remains mostly as solid salts on land (in rock and soil minerals). Weathering of rocks carries these phosphates to terrestrial habitats. Plants absorb phosphates from the soil and then are consumed by herbivores that in turn may be consumed by carnivores. After death, the animal or plant decays to return phosphates to the soil. Runoff may carry them back to the ocean to be reincorporated into rock. The processes that move them through the soil or ocean are very slow, making the phosphorus cycle over all one of the slowest biogeochemical cycles.

Phosphorus is an essential nutrient for plants and animals in the form of ions and is a limiting nutrient for aquatic organisms. Human interference in the phosphorus cycle occurs by over use or careless use of phosphorus fertilizers, which results in phosphorus as pollutants in bodies of water with eutrophication consequences.

Sulphur Cycle

The sulphur cycle is similar to the nitrogen cycle in that microbial processes control which form of sulphur appears in the natural environment. The most oxidized form of sulphur is sulphate which is the form most readily assimilated by plants to be reduced into organic sulphur. Sulphur is a key component in proteins, amino acids and B vitamins of plants and animals. Once plants and animals die, bacteria turn organic sulphur into reduced sulphide (H_2S) and then into sulphate to complete the cycle.

Source: Han D. Concise Environmental Engineering[M]. Download free ebooks at bookboon.com.

Words and Phrases

topography [tə'pɒgrəfi] *n.* 地貌；地形学；地形测量学

decomposition [ˌdiːkɒmpə'zɪʃn] *n.* 分解；腐烂

succession [sək'seʃn] *n.* 连续；继位；继承权；[生态] 演替

calorific [ˌkælə'rɪfɪk] *adj.* 发热的，生热的

trophic ['trəʊfɪk] *adj.* 营养的

autotroph ['ɔːtətrəʊf] *n.* 自养生物

heterotroph ['hetərətrəʊf] *n.* 异养生物

carnivore ['kɑːnɪvɔː(r)] *n.* 食肉动物，食虫植物

detritivore ['detrɪtɪvɔː] *n.* 食碎屑者；腐食性生物

fixation [fɪk'seɪʃn] *n.* 固定；定居

photoautotroph [ˌfəʊtəʊ'ɔːtətrəʊf] *n.* 光自养生物

digestion [daɪ'dʒestʃən] *n.* 消化，细菌分解

respiration [ˌrespə'reɪʃn] *n.* 呼吸；呼吸作用

excretion [ɪk'skriːʃn] *n.*（动植物的）排泄，排泄物

regurgitation [rɪˌɡɜːdʒɪ'teɪʃn] *n.* 回流；反刍

residence ['rezɪdəns] **time** 停留时间；残留时间

lithosphere ['lɪθəsfɪə(r)] *n.* 岩石圈，岩石层

combustion [kəm'bʌstʃən] *n.* 燃烧；燃烧过程

biosynthesis [ˌbaɪəʊ'sɪnθɪsɪs] *n.* 生物

合成

ammonification [əmɒnɪfɪ'keɪʃn] n. [化学] 氨化；加氨

nitrification [ˌnaɪtrəfɪ'keɪʃən] n. 氮化合，氮饱和，[化学] 硝化作用

sequestered [sɪ'kwestəd] adj. 隐退的；幽静的；扣押的；偏僻的

denitrification [diːˌnaɪtrɪfɪ'keɪʃn] n. 脱氮；[化学] 反硝化作用

phosphate ['fɒsfeɪt] n. 磷酸盐；磷肥

terrestrial [tə'restriəl] adj. 陆地的，陆生的

eutrophication [juːtrəfɪ'keɪʃn] n. 富营养化；超营养作用

sulphate ['sʌlfeɪt] n. 硫酸酯；[化学] 硫酸盐；vt. 使与硫酸化合；vi. 硫酸盐化

sulphide ['sʌlfaɪd] n. 硫化物

tertiary consumer 三级消费者

body mass 生物量

aerobic respiration 有氧呼吸；需氧呼吸

active carbon 活性炭；活性碳

inert carbon 惰性碳

Questions

1. What is an ecosystem?
2. List the factors that control an ecosystem.
3. What are the two things that living organisms require from the environment?
4. What limits typical food chains to only four to six links?
5. List the main reservoirs of carbon, and sort them by content.

Sentence-making

1. producers, consumers, decomposers
2. autotroph, bacteria, nutrition
3. heterotrophy, detritivore, digestion
4. ammonification, nitrification, denitrification
5. eutrophication, phosphate, anthropogenic

Part B Extensive Reading

Ecology and Environmentalism

Our concern over the condition of the natural environment has led to the introduction of a new concept, of "environmental quality", which can be measured against defined parameters. It is also possible, though much more difficult, to determine the quality of a natural habitat in terms of the species it supports and to measure any deterioration as the loss of species.

These are matters that can be evaluated scientifically, in so far as they can be measured, but not everything can be measured so easily. We know, for example, that in many parts of the tropics primary forests are being cleared, but although satellites monitor the affected areas it is difficult to form accurate estimates of the rate at which clearance is proceeding, mainly because different people classify forests in different ways and draw different boundaries to them. The United Nations Environment Programme (UNEP) has pointed out that between 1923 and 1985 there were at least 23 separate estimates of the total area of closed forest in the world, ranging from 23.9 to 60.5 million km^2. The estimate of UNEP prefers suggests that in pre-agricultural times there was a total of 12.77 million km^2 of tropical closed forests and that by 1970 this had been reduced by 0.48 percent, to 12.29 million km^2, and that the total area of forests of all kinds declined by 7.01 percent, from 46.28 to 39.27 million km^2, over the same period. Edward O. Wilson, on the other hand, has written that in 1989 the total area of rain forests was decreasing by 1.8 percent a year. A rain forest is one in which the annual rainfall exceeds 2540 mm; most occur in the tropics, but there are also temperate rain forests. Similar differences occur in estimates of the extent of land degradation through erosion and the spread of deserts (called "desertification"). Before we can devise appropriate responses to these examples of environmental deterioration we have to find some ways of reconciling the varying estimates of their extent. After all, it is impossible to address a problem unless we can agree on its extent.

Even when quantities can be measured with reasonable precision controversy may attend interpretations of the measurements. We can know the concentration of each substance present in air, water, soil, or food in a particular place at a particular time. If certain of those substances are not ordinarily present and could be harmful to living organisms we can call them "pollutants", and if they have been introduced as a consequence of human activities, rather than as a result of a natural process such as volcanism, we can seek to prevent further introduction of them in the future. This may seem simple enough, but remember that someone has to pay for the measurement: workers need wages, and equipment and materials must be bought. Reducing pollution is usually inconvenient and costly, so before taking action, again we need to determine the seriousness of the problem. The mere presence of a pollutant does not imply harm, even when the pollutant is known to be toxic. Injury will occur only if susceptible organisms are exposed to more than a threshold dose, and where large numbers of very different species of plants, animals, and microorganisms are present this threshold is not easily calculated.

Nor is it easy to calculate thresholds for human exposure, because only large populations can be used for the epidemiological studies that will demonstrate effects, and small changes cannot always be separated statistically from natural fluctuations (Epidemiology is the study of the incidence, distribution, and control of illness in a human population.). It has been estimated that over several decades the 1986 accident at the Chernobyl nuclear reactor may lead to a 0.03 percent increase in radiation-induced cancer deaths in the former Soviet Union and a 0.01 percent increase in the world as a whole, that will not be detectable against the natural variations in the incidence of cancer from year to year.

Where there is doubt, prudence may suggest we set thresholds very low, and in practice this is what happens. With certain pesticide residues in food, for example, the EU operates a standard of "surrogate zero" by setting limits lower than the minimum quantity that can be detected.

Where the statistical evaluation of risk is unavoidably imprecise yet remedial action seems intuitively desirable, decisions cannot be based solely on scientific evidence and are bound to be more or less controversial. Since decisions of any kind are necessarily political, and will be argued this way and that, people will take sides and issues will tend to become polarized.

At this point, environmental science gives way to environmental campaigning, or environmentalism, and political campaigns are managed by those activists best able to publicize their opinion. In their efforts to attract public attention and support, spokespersons are likely to be drawn into oversimplifying complex, technical issues which, indeed, they may not fully understand, and to exaggerate hazards for the sake of dramatic effect.

Ecology is a scientific discipline devoted to the study of relationships among members of living communities and between those communities and their abiotic environment. Intrinsically it has little to do with campaigning for the preservation of environmental quality, although individual ecologists often contribute their professional expertise to such campaigns and, of course, their services are sought whenever the environmental consequences of a proposed change in land use are assessed.

"Ecology", then, is at one and the same time a scientific discipline and a political, at times almost religious, philosophy which inspires a popular movement and "green" political parties in many countries. As a philosophy, it no longer demands piecemeal reform to achieve environmental amelioration, but calls for the radical restructuring of society and its economic base. The two meanings attached to the word are now quite distinct and it is important not to confuse them. When people say a particular activity or way of life is "ecologically sound" they are making a political statement, not a scientific one, even though they may be correct in supposing the behaviour they approve to have less adverse effect on human health or the welfare of other species than its alternatives. "Ecologically sound" implies a moral judgement that has no place in scientific argument; to a scientist the phrase is meaningless.

This is not to denigrate those who use the word "ecology" in one sense or the other, simply to point out that the meanings are distinct and our attitudes to the environment are shaped by historical, social and economic forces. They are not derived wholly from a scientific description of the environment or understanding of how it works. The nuclear power industry, for example, is opposed on ecological grounds, but there is no evidence that it has ever caused the slightest injury to non-humans, apart from vegetation around the Chernobyl complex following the accident there, and its adverse effects on human health are extremely small, especially when compared with those resulting from other methods of power generation; indeed, it is extremely unlikely that the correct routine operation of a nuclear power plant has any harmful effect at all, on humans or non-humans. The anti-nuclear wing of the environmental movement is highly influential and has done much to erode public confidence in the industry, but whether this is environmentally beneficial is open to debate, to say the least. In contrast, on those occasions when scientists and campaigners collaborate, say in devising (scientifically) the best way to manage an area in order to maximize its value as natural habitat then campaigning (politically) to have the area protected from inappropriate development, they can achieve their useful and practicable goal. While it is certainly true that some ecological (i.e. environmentalist) campaigns owe little to ecology (the science), others, though not necessarily the most populist, are scientifically well informed. It is also true that if we confine our interest to the acquisition of an abstract understanding of the way the world is, that understanding will be of limited practical value. If damage to the environment is to be avoided or past damage remedied, scientific understanding must be applied and this is possible only through political processes.

Source: Allaby M. Basics of Environmental Science, 2nd Edition[M].The Taylor & Francis e-Library, 2002.

Words and Phrases

environmentalism [inˌvaɪrən'mentəlɪzəm]　*n.* 环境保护论
[例句] The problem for China, though, is that environmentalism is a bottom-up movement in the rest of the world.

desertification [dɪˌzɜːtɪfɪ'keɪʃn]　*n.* (土壤)荒漠化
[例句] More than one quarter of China is either covered by desert or is land that is suffering desertification.

reconcile ['rekənsaɪl]　*vt.* 使和解；调停；使安心；使一致
[例句] The teacher in charge of our class had to reconcile disputes among the students.

precision [prɪ'sɪʒn]　*n.* 精度，[数] 精密度；精确；*adj.* 精密的，精确的
[例句] If you have numeric data, keep the level of precision appropriate.

susceptible [sə'septəbl]　*adj.* 易受影响的；易感动的
[例句] Young people are the most susceptible to advertisements.

threshold ['θreʃhəʊld]　*n.* 入口；门槛；开始；极限；临界值
[例句] If this threshold is exceeded, an alert is sent to the notification list.

epidemiological [ˌepɪˌdiːmɪə'lɒdʒɪkl]　*adj.* 流行病学的
[例句] They carried out clinical and epidemiological examinations of suspected cases in hospitals and at home, and collected clinical samples.

pesticide ['pestɪsaɪd]　*n.* 杀虫剂
[例句] This pesticide is diluted with water and applied directly to the fields.

maximize ['mæksɪmaɪz]　*vt.* 使增加到最大限度
[例句] The only catch is you have to know how to use it and how to maximize its potential once you have it.

minimum quantity　最低数量　　　　**sparrow hawk**　食雀鹰；鹞
remedial action　补救行动，矫正措施　　**environmental amelioration**　环境治理
abiotic environment　非生物环境　　　**radical restructuring**　根本重建

Questions

1. What is ecology?

2. What are pollutants?

3. Why is it difficult to form accurate estimates of the rate at which clearance is proceeding?

4. Why does not the mere presence of a pollutant imply harm, even when the

pollutant is known to be toxic?

Part C Dialogue

Dialogue 1

Anna: I was reading the newspaper the other day and one of the news about the American action movie "The Expendables 2" caught my attention.
Bob: What did it say?
Anna: The dramatic team was being accused of destroying the Bulgaria ecological environment. The number of local bats has significantly dropped from over 30 thousand to less than 9 thousand.
Bob: Really! I guess the team must have been shooting the film in the cave those bats live in.
Anna: Yeah, exactly. Experts said that the noise and light disturbed the bats so severely that they couldn't sleep normally at night. You know what, I think such abnormal changes in species population size could trigger unexpected problems.
Bob: What do you mean?
Anna: Remember in the 1950s, Chinese farmers hunted sparrows because they eat grains? An insect pest broke out after that.
Bob: Oh, I see. It is because people accidentally wiped out those insects' natural enemy. Anna, I'm a bit worried now. Industrialization and farming are causing much bigger problems than shooting films. Wetland, forest, species are losing their habitation at a growing rate. I'm a bit pessimistic about the situation.
Anna: Yeah, if the ecosystem fails to function normally, we would definitely suffer. Issues about climate change and pollution are all over the newspaper.
Bob: Right. We have seen strange weather conditions this year, like the flood in India, the hail in Australia, the snow storm in Europe…
Anna: Yeah, my friend Ella was trapped in the London airport for three days, poor girl.
Bob: Anna, guess what I think is the top one miserable natural disaster?
Anna: Ah… the super-hot summer in India? 50 degrees, I would definitely faint.
Bob: That's a nice try. Well, I would rank the Fukushima earthquake first due to the follow-up nuclear leakage. While we consider ourselves capable of reshaping the natural environment to satisfy our needs, we end up hurt by the technology we use. Don't you think so?
Anna: I totally agree with you, Bob. I think the nature has a perfect system that no

artificial design could ever match. For us to live better, a balance must be stuck between development and environment protection.

Bob: Fortunately, we have seen actions by the government to deal with these ecological issues.

Anna: Let's just wish more progresses could be made in the future.

Source: http://www.zybang.com/question/a99427bd9609722fab062fb7d06b3cb7.html

Dialogue 2

Steve McCormick: president of the Gordon and Betty Moore Foundation, and CEO at The Nature Conservancy. As president and a trustee of the Gordon and Betty Moore Foundation, Steve leads the foundation's work to turn bold ideas into enduring impact in environmental conservation, science and patient care.

Mark Tercek: What do you see as the biggest environmental challenges we face — and the best ways to address them?

Steve McCormick: Well, let me start by saying that the biggest reality we need to confront is that, despite all the great conservation work that's been done over the past 50 years, we haven't staunched the rate of loss of biodiversity. Extinction rates have actually increased in the 30 years I've worked in conservation. And while some major habitat types, like temperate forests, have been fairly well conserved, most are disappearing at accelerated rates.

The biggest challenge we have to face, therefore, is that the drivers of these changes derive from fundamental human needs and desires. While I am passionate about the importance of creating protected areas, I'm convinced that the only way we will secure conservation at a globally meaningful scale is to work on system change, especially in market systems by reflecting the true costs of natural capital in its various forms.

Mark Tercek: What else could the environmental movement do to better address these challenges?

Steve McCormick: Frankly, I don't see a "movement" at all. I see individual organizations doing good projects, but the aggregate result of those projects isn't adding up to mission success for any of the organizations. I would really like to see more collective effort towards really ambitious shared goals — goals that aim to reduce rates of loss of biological diversity.

Mark Tercek: Yes, I agree. We're making some progress here — including our merger with Rare and the Science for Nature and People (SNAP) initiative, our new coalition with Wildlife Conservation Society and the National Center for Ecological Analysis and Synthesis — but we still have a considerable way to go. What other advice do you have for how we can make the biggest difference in our mission?

Steve McCormick: Two things come immediately to mind. First, align everything around mission and strategy, meaning be willing to make tough trade-offs. I see a lot of organizations layering in new programs and activities without getting out of anything. Related to that, be absolutely clear on what constitutes mission success and be disciplined in assessing what will optimize achieving those goals.

Second, I strongly encourage conservation CEOs to find opportunities to work towards collective impact. I see from my current role how the conservation organizations could have much greater impact by working in complementarity. I have to confess, foundations could also do the same. It's something I'm committed to here.

Mark Tercek: How is Moore trying to achieve "collective impact"? What are the best opportunities? And what would you do if you were in my shoes at The Nature Conservancy or at another NGO?

Steve McCormick: We have several collective efforts underway. One example is a joint venture — the Science Philanthropy Alliance — with six other large foundations that, like us, support basic research, and about a dozen research universities. The goal is to increase private funding for research by $1 billion over the next five years. In our Patient Care Program we are creating a coalition of four academic medical centers and three other foundations to create a scalable IT platform designed to dramatically decrease preventable harms in the ICU setting. And a good, recent example is the mobilization of key stakeholders in stopping the Pebble Mine at Bristol Bay in Alaska, led by a former TNC'er, Erin Dovichin.

As for what you and your colleague CEOs in conservation could do is sharing staff resources on a major issue — say energy — so that each NGO doesn't create duplicative staffing and approaches. Get together and create a shared picture of a pragmatic energy future, then deploy the best, complementary staff from each organization. And give the joint effort sufficient decision-making authority and resources. Another opportunity is working with corporations on sustainability. Recognize that we need a common methodology around value of nature in business decisions, instead of individual one-off relationships with individual companies. The multiple approaches and models are creating confusion, and therefore lack of systemic, industry-wide uptake.

Mark Tercek: What else excites you about your work at Moore Foundation?

Steve McCormick: Foundations, especially large foundations like ours, with the size and reliability of our financial resources, have the rare capability of trying to effect systemic change. We can deploy patient capital; we can support unorthodox but potentially breakthrough ideas; we can strike quickly to create momentum; and we can help mobilize collective action. It's really very exciting.

Mark Tercek: What are some of the most exciting opportunities in this regard that you are pursuing now?

Steve McCormick: Well, all of our Environment Program Initiatives — Andes-Amazon; Marine Conservation; Wild Salmon Ecosystems — are examples of "patient capital;" we've been involved with them for over ten years. As we move into addressing sustainability we will take the same long view. A good example of "quick strike" was our support of EDF's imaginative habitat exchange approach to addressing the conflicts that arise when a species is listed as endangered. When we realized they had an opportunity to move other key players, but needed to act fast, we got a $1 million contribution to them immediately.

Mark Tercek: I argue in my book Nature's Fortune that focusing on nature as an investment opportunity can get people who may have been viewed as opponents of the environmental movement on our side, provide a source of capital and an opportunity to scale up. What risks and opportunities do you see in this approach?

Steve McCormick: The problem is that markets don't reflect the true costs of natural capital. The concept of externalities isn't new, but until markets incorporate these costs we'll continue to see investment decisions distorted by artificially high returns. So-called impact investing won't be of sufficient scale to have sufficient influence on the overexploitation of natural capital.

Mark Tercek: Looking back, what is something that you've been wrong about in the past? How has that changed your thinking today?

Steve McCormick: Gosh, I've been wrong about so many things. Frankly, one of the things that I was "wrongest" about was trying to bring about substantial change in TNC. I tried to change too much, too fast. What I learned is just how profoundly hard it is to change big, established organizations.

Source: http://www.huffingtonpost.com/mark-tercek/qa-with-steve-mccormick-d_b_4031374.html

Part D Supplementary Vocabulary

Ecology

aciculignosa 针叶木本群落
agribiodiversity 农业生物多样性
agrometeorology 农业气象学
air quality index 大气质量指数
alkali land 碱地
aquatic plant 水生植物
arid land ecosystem 干旱地区生态系统
artificial environment 人工环境
biocoenosis 生物群落
bioconcentration 生物浓缩
biogeochemical cycle 生物地球化学循环
biological enrichment 生物富集
biological invasion 生物入侵
biomagnification 生物放大
biomass 生物量
biome 生物群落区
biosphere 生物圈
canopy density 郁闭度
characteristic species 特征种
chromosomal mutation 染色体突变
circle of vegetation 植被圈
climax community 顶级群落
coevolution 协同进化
cold acclimation 低温驯化
commensalism 偏利作用
compartmentalization 隔离作用
conjunctive symbiosis 共生
contagious distribution 集群分布
convergent adaptation 趋同适应
courtship behavior 求偶行为
depression effects 抑制效应

dominant species 建群种，优势种
drought-deciduous species 干旱落叶植物
ecological amplitute 生态幅
ecological density 生态密度
ecological engineering 生态工程
ecological extinction 生态灭绝
ecological invasion 生态入侵
ecological pyramid 生态金字塔(生态锥体)
ecotype 生态型
edge effect 边缘效应
emersiherbosa 湿生草本群落
energy flow chart 能流图
energy flow structure 能流结构
environment hormone 环境激素
environmental capacity 环境承载力
etiolating phenomenon 黄化现象
eutrophication 富营养化
evergreen hardwood forest 常绿硬木林
ex-situ conservation 迁地保护
exotic species / alien species 外来种
exponential growth 指数增长
feed back 反馈作用
food chain structure 食物链结构
freshwater ecosystem 淡水生态系统
frigorideserta 冻荒漠群落
functional groups 功能群
gene pool 基因库
genetic deterioration 遗传衰退
genetic drift 遗传漂变

genotype 基因型
greenhouse effect 温室效应
gross primary production 总初级生产力
halophyte 盐土植物
humus 腐殖质
hydrarch succession 水生演替
hybrid speciation 杂交成种
hydrosphere 水圈
in situ conservation 就地保护
interspecies competition 种间竞争
intraspecific competition 种内竞争
introduced species 引进种
invasive species 侵入种
isozyme 同工酶
landscape ecology 景观生态学
laurilignosa 常绿阔叶林
law of tolerance 耐受性法则
leached layer 淋溶层
lithosphere 岩石圈
logistic growth 逻辑斯蒂增长
mangrove forest 红树林
marine ecosystem 海洋生态系统
mark-recapture 标记-重补法
monoclimax theory 单元演替顶级
monsoon circulation 季风环流
monsoon forest 季雨林/季风林
mutation 突变
mutualism 互利共生
natality 出生率
native species/indigenous species 本地种/土著种
negative feedback 负反馈
niche overlap 生态位重叠
ocean-current 洋流
plankton 浮游生物
population ecology 种群生态学
positive feedback 正反馈
predator 捕食者
primary succession 原生演替
protogynous hermaphroditism 雌雄同体
pyramid of energy 能量金字塔
radiation adaptation 趋异适应
richness 丰度
secondary production 次级生产力
secondary succession 次生演替
sexual reproduction 有性生殖
species diversity 物种多样性
synecology 群落生态学
trophic level 营养级
vegetative propagation 营养生殖
xerarch succession 旱生演替
zone of emergent vegetation 挺水植物带
zooplankton 浮游动物

Part E   Supplementary Knowledge

无机化合物命名法

1. Monatomic Ions 单原子离子的命名

- 阳离子

元素名＋ion 中英文同

Na^+: sodium ion 钠离子

Mg^+: magnesium ion 镁离子
过渡元素，有多种离子态时
Mn^{2+}: manganese II ion 二价锰离子
Fe^{2+}: iron II ion 读法：iron two
Fe^{3+}: iron III ion 读法：iron three
- 阴离子

中文：元素名+离子，如：氯离子
英文：元素名（**ide**）ion
Cl^-: chlorine → chlor**ide** ion
S^{2-}: sulfur → sulf**ide** ion
P^{3-}: phosphorus → phosph**ide** ion

Number	Prefix
1	mono-
2	di-
3	tri-
4	tetra-
5	penta-
6	hexa-
7	hepta-
8	octa-
9	nona-
10	deca-
12	dodeca-
14	tetradeca-

- 俗名法

较低价数者，尾端为 **ous**（亚）；较高价数者，尾端为 **ic**
例如 Fe^{2+}: **ferrous** ion 亚铁离子； Fe^{3+}: **ferric** ion
　　　Cr^{3+}: **chromic** 三价铬；Cr^{2+}: **chromous** 二价铬、亚铬

2. Polyatomic Ions 多原子离子（根）的命名

- 含氧原子：字尾为 **ate** 和 **ite**；ite 比 ate 少一个氧原子
  但不同 ate、ite 离子物质的价数（电荷数）不一定相同
  例　NO_3^-: nit**rate** ion 硝酸根　　NO_2^-: nit**rite** ion 亚硝酸根
  　　SO_4^{2-}: sulf**ate** ion 硫酸根　　SO_3^{2-}: sulf**ite** ion 亚硫酸根

- 前缀 **bi** 代表氢，而非"2"
  bicarbonate ion HCO_3^- 碳酸氢　　**bi**sulfate ion HSO_4^- 硫酸氢

- 前缀 **per** 代表较 ate 多一个氧（过），**hyp**-代表较 ite 少一个氧（次）
  ClO_3^-: chl**orate** 氯酸根　　　　ClO_2^-: chl**orite** 亚氯酸根
  ClO_4^-: **per**chlorate 过氯酸根　　ClO^-: **hyp**ochlorite 次氯酸根

- 前缀 **thio**：一个氧被硫取代（硫代）
  SO_4^{2-}: sulfate ion 硫酸根→　$S_2O_3^{2-}$: **thio**sulfate ion 硫代硫酸根
  OCN^-: cyanate 氰酸根 →　SCN^-: **thio**cyanate 硫代氰酸根

3. 化合物的命名

- 英文命名法：正价元素名称 + 负价元素名称的词干 + -ide

CaO: calcium oxide　　　　　　CO_2: carbon dioxide
P_2O_5: diphosphorus pentoxide　　SF_6: sulfur hexafluoride
CO: carbon monoxide　　　　　FeO: ferrous oxide
Fe_2O_3: ferric oxide　　　　　　V_2O_5: divanadium pentoxide

4. 酸的命名

- 常用俗名，而不用其正规命名，如 HNO_3 硝酸为 nitric acid，而非 hydrogen nitrate。
- 酸的命名：酸根离子中非氧元素名称的词干 + -ic acid

如果某元素能形成一种以上的含氧酸，则按以下规则：

    a. 高(过)＊酸：per- + 酸根离子中非氧元素名称的词干 + -ic acid，如 perchloric acid

    b. ＊酸：酸根离子中非氧元素名称的词干 + - ic acid ，如 chloric acid, nitric acid

    c. 亚＊酸：酸根离子中非氧元素名称的词干 + -ous acid，如 nitrous acid

    d. 次＊酸：hypo- + 酸根离子中非氧元素名称的词干 + -ous acid，如 hypochlorous acid（从 a 到 d 含氧原子数依次递减）

    e. 偏＊酸：meta- + 酸根离子中非氧元素名称的词干 + -ic acid

    f. 焦＊酸：pyro- + 酸根离子中非氧元素名称的词干 + -ic acid

5. 碱的命名

- 碱的命名：元素名称 + hydroxide

    例：NaOH：sodium hydroxide

    KOH：potassium hydroxide

- 如果某元素能形成一种以上的阳离子，则使用斯托克数字（Stock number）来表示其所带电荷（只形成一种阳离子的不必用）。

6. 盐的命名

- 英：cation-anion   → 中：阴化阳

    英：NaCl   → 中：氯化钠

    英：由前而后   → 中：由阴而阳（数量必须注明）

英文命名法：不带"ion"的阳离子名称 + 不带"ion"的阴离子名称，

例如：NaCl：sodium chloride

- 除非可以从化学价简单判断，多价离子须注明。

    例：NaCl：sodium chloride

        CuCl：copper(I) chloride

        $CuCl_2$：copper(II) chloride

        $CuSO_4$：copper(II) sulfate

        iron chloride   × 错误表示法，无法表示种类

        iron chlorides   √ 泛指 $FeCl_2$、$FeCl_3$ 的统称

7. Hydrates 水合物的命名

- 命名：无水盐 + 希腊数字 + hydrate

例：$CuSO_4 \cdot 5H_2O$：copper(II) sulfate pentahydrate

Reference: https://wenku.baidu.com/view/577cf0a30029bd64783e2caa.html?qq-pf-to=pcqq.c2c
https://wenku.baidu.com/view/577cf0a30029bd64783e2caa.html

Unit 3　Energy and Climate Change

Part A　Intensive Reading

Overview of Climate Change Science

Overview

Earth's climate is changing. Multiple lines of evidence shows changes in our weather, oceans, ecosystems, and more. Natural causes alone cannot explain all of these changes. Human activities are contributing to climate change, primarily by releasing billions of tons of carbon dioxide (CO_2) and other heat-trapping gases, known as greenhouse gases, into the atmosphere every year. Climate changes will continue into the future. The more greenhouse gases we emit, the larger future climate changes will be.

Earth's climate is changing

The global average temperature has increased by more than 1.5°F since the late 1800s. Some regions of the world have warmed by more than twice this amount. The buildup of greenhouse gases in our atmosphere and the warming of the planet are responsible for other changes, such as:

- Changes in temperature and precipitation patterns
- Increases in ocean temperature, sea level, and acidity
- Melting of glaciers and sea ice
- Changes in the frequency, intensity, and duration of extreme weather events
- Shifts in ecosystem characteristics, like the length of the growing season, timing of flower blooms, and migration of birds
- Increasing effects on human health and well-being

Natural causes alone cannot explain recent changes

Natural processes such as changes in the sun's energy, shifts in ocean currents, and others affect earth's climate. However, they do not explain the warming that we have observed over the last half-century.

Human causes can explain these changes

Most of the warming of the past half century has been caused by human emissions of greenhouse gases. Greenhouse gases come from a variety of human activities, including:

burning fossil fuels for heat and energy, clearing forests, fertilizing crops, storing waste in landfills, raising livestock, and producing some kinds of industrial products.

Models that account only for the effects of natural processes are not able to explain the warming observed over the past century. Models that also account for the greenhouse gases emitted by humans are able to explain this warming (Figure 1).

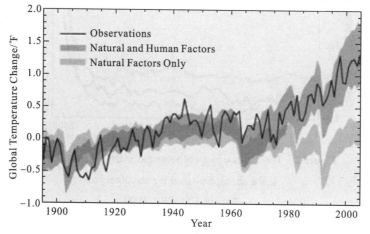

Figure 1 Separating human and natural influences on climate change

Source: U.S. National Climate Assessment (2014)

The extent of future climate change depends on us

The extent of the change will depend on how much, and how quickly, we can reduce greenhouse gas emissions. During the 21st century, global warming is projected to continue and climate changes are likely to intensify (Figure 2). Scientists have used climate models to project different aspects of future climate, including temperature, precipitation, snow and ice, ocean level, and ocean acidity. Depending on future emissions of greenhouse gases and how the climate responds, average global temperatures are projected to increase worldwide by 0.5°F to 8.6°F by 2100, with a likely increase of at least 2.0°F for all scenarios except the one representing the most aggressive mitigation of greenhouse gas emissions.

Increases in concentrations of these gases since 1750 are due to human activities in the industrial era. Concentration units are parts per million (ppm) or parts per billion (ppb), indicating the number of molecules of the greenhouse gas per million or billion molecules of air.

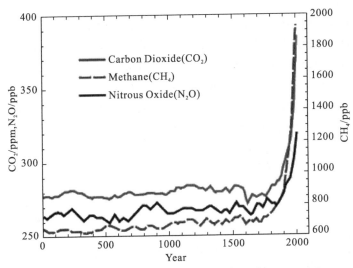

Figure 2 The increase in greenhouse gas (GHG) concentrations in the atmosphere over the last 2000 years

Source: USGCRP, 2009

Climate change impacts our health, environment, and economy

Climate change affects our environment and natural resources, and impacts our way of life in many ways. For example:

- Warmer temperatures increase the frequency, intensity, and duration of heat waves, which can pose health risks, particularly for young children and the elderly.
- Climate change can also impact human health by worsening air and water quality, increasing the spread of certain diseases, and altering the frequency or intensity of extreme weather events.
- Rising sea levels threaten coastal communities and ecosystems.
- Changes in the patterns and amount of rainfall, as well as changes in the timing and amount of stream flow, can affect water supplies and water quality and the production of hydroelectricity.
- Changing ecosystems influence geographic ranges of many plant and animal species and the timing of their lifecycle events, such as migration and reproduction.
- Increases in the frequency and intensity of extreme weather events, such as heat waves, droughts, and floods, can increase losses to property, cause costly disruptions to society, and reduce the availability and affordability of insurance.

Greenhouse gas emissions are not the only way that people can change the climate.

Activities such as agriculture or road construction can change the reflectivity of earth's surface, leading to local warming or cooling. This effect is observed in urban centers, which are often warmer than surrounding, less populated areas. Emissions of small particles, known as aerosols, into the air can also lead to reflection or absorption of the sun's energy.

Source: https://19january2017snapshot.epa.gov/climate-change-science/overview-climate-change-science_.html

Words and Phrases

precipitation [prɪˌsɪpɪ'teɪʃn] *n.* [化学] 沉淀，[化学] 沉淀物；冰雹；坠落
glacier ['ɡlæsɪə(r)] *n.* 冰河，冰川
landfill ['lændfɪl] *n.* 垃圾填埋地；垃圾堆
rainfall ['reɪnfɔːl] *n.* 降雨；降雨量
hydroelectricity [ˌhaɪdrɔɪ'lɛktrɪsəti] *n.* 水力电气
particle ['pɑːtɪkl] *n.* 颗粒；[物] 质点；极小量
aerosol ['eərəsɒl] *n.* [物化] 气溶胶；气雾剂；喷雾器；浮质
heat-trapping gas 吸热气体
greenhouse gas 温室气体
average surface temperature 表面平均温度
precipitation pattern 降水模式

extreme weather event 极端天气事件
ecosystem characteristic 生态系统特征
migration of birds 鸟类迁徙
shift in ocean currents 洋流的变化
fossil fuel 化石燃料
clearing forest 砍伐森林
fertilizing crop 施肥的作物
raising livestock 饲养牲畜
heat wave 热浪
stream flow 流量
road construction 道路建设
stormwater system 雨水系统
water conservation program 水资源保护项目
warning system 预警系统
emergency preparation 应急准备
response strategy 反应策略

Questions

1. Is climate change the same thing as global warming?
2. What are the most visible signs of climate change?
3. Has the earth climate changed before? What's different from today's climate?

Sentence-making

1. water conservation, stormwater systems, emergency preparation
2. extreme weather events, ecosystem characteristics, migration of birds

3. greenhouse gases, heat-trapping gases, fossil fuels

4. particle, aerosol, adsorption

5. precipitation, acid, sulfur

Part B   Extensive Reading

Energy, Climate Change and Environment

Policies and actions to "unlock" high-emissions assets

Meeting the challenge of climate change is not only about channelling new investments toward clean energy, but also addressing high-emissions assets that are already in place. Long-lived infrastructure can create path dependence in energy systems and the potential for lock-in. Staying on track to limit temperature rise to below two degrees Celsius requires a transition away from these assets at faster rates than natural infrastructure replacement would dictate (i.e. before the end of their economic lifetimes). Current assets could be seen as "locked in", but they can also be "unlocked" through policy intervention.

High and rising carbon prices could drive changes in infrastructure; however, given the low prices in most current carbon pricing systems today, alternative policy options need to be explored to unlock high-emissions infrastructure. The context of coal plants, one of the largest sources of energy sector greenhouse gas (GHG) emissions, provides useful insights.

In choosing policy options to unlock existing infrastructure, careful attention needs to be paid to not undermining long-term outcomes. For example, it is critical that any early retirements will be replaced by clean generation. Equally, policies to drive deployment of clean generation need to be complemented by policies that address fossil fuel emissions, in order to avoid unintended consequences such as the mothballing of gas plants instead of coal in Europe. Moreover, energy security should always be a priority to produce sustainable actions from an energy perspective: early retirements need to be matched with new supply or energy efficiency gains to keep reserve margins at acceptable levels.

The new landscape of emissions trading systems

Emissions trading systems (ETSs) are enjoying somewhat of a resurgence around the world. As a form of carbon pricing, ETSs represent effective and low-cost policy responses to climate change. Beginning with the European Union Emissions Trading Scheme in 2005, which remains the largest system, current or planned systems now exist in all corners of the globe. Since 2013 the world has seen a rise in ETS implementation, with new or expanded systems in China, California, Québec, Kazakhstan and Switzerland. The Northeast United States, New Zealand, and Tokyo are other examples, and there are more under preparation: Republic of Korea has passed legislation to begin emissions trading by 2015, and India, Chile, Brazil, Thailand and Mexico are in various stages of consideration and development of ETSs. While it is clear that support for carbon pricing and emissions trading is not universal, it is difficult to ignore the trend of expansion.

Key lessons can be drawn from recent ETS experiences:
- Improved integration of ETSs and complementary energy policies can ensure each set of policies meeting its respective objectives.
- Measures can be taken to enhance ETS resilience and flexibility within changing economic conditions.
- ETS design must consider changing political contexts and public perceptions given that real as well as perceived impacts determine policy success.
- ETSs may be implemented in highly regulated electricity systems, though additional measures may be needed to ensure propagation of the carbon price signal.
- Compensating those groups affected by rising electricity prices (driven by the carbon price) may achieve better outcomes than preventing the price rise.

Lastly, although the role of ETSs within an international climate change agreement remains uncertain despite their global expansion, the United Nations Framework Convention on Climate Change (UNFCCC) process has important potential functions to play. The UNFCCC process can help balance, on one hand, flexibility for countries to develop their own market-based approaches to GHG reductions with, on the other, the need to establish common international rules and standards to build trust and credibility.

Energy metrics: A useful tool for tracking decarbonization progress

While GHG emission reduction goals are an essential component of decarbonization, specific energy sector metrics can provide deeper insight into the underlying drivers of change, and can track interventions with long-term as well as short-term impacts. Energy sector policies and actions that reduce GHG emission may be motivated primarily by wider benefits such as energy security, building experience with new technologies, cutting air pollution, or reducing energy bills, with GHG emission reductions as a secondary benefit.

There are many reasons why countries may be motivated to use energy sector goals and metrics to support GHG emission reductions:
- Goals based on energy sector metrics can link more directly to policies under government control (e.g. renewable portfolio standards). They may consequently be easier to adopt, as outcomes are more easily influenced by policy and decision makers can have more confidence in delivery.
- Clean energy policies are implemented for a wide range of reasons and often have multiple benefits, of which emission reductions are only one; accordingly, energy sector metrics may better reflect these objectives.
- Discussions toward the new 2015 agreement seek to frame climate action as an opportunity to be seized rather than a burden to be shared. Energy metrics can potentially help change the discourse around climate goals.
- Alternative metrics can help to target the long-term transformation that is needed to complement short-term goals (e.g. to prevent lock-in of high-emission infrastructure or support the development of key clean technologies).

The use of energy sector metrics in addition to GHG goals could be helpful within and outside the UNFCCC process to help drive the energy sector actions needed for decarbonization. The use of metrics that better reflect the various goals of parties could help them build support for climate policy.

The air pollution-GHG emissions nexus: Implications for the energy sector

The energy sector is the greatest contributor to heat-trapping GHG emissions through the combustion of fossil fuels. Fossil fuel combustion also causes air pollution, which poses increasingly pressing problems around the world as public health and economic damages continue to accrue in countries at all levels of development. This presents critical challenges for the production and use of energy, which is central to economic growth and development.

However, opportunities are available to "co-manage" these challenges at the air pollution-GHG emissions nexus in a variety of contexts. This is especially important since the interplay between air pollution control and GHG emissions abatement may not always be positive. With this in mind, a special focus in this year's publication is on the linkages between air pollution control and GHG emissions:
- GHG co-benefits of air quality controls of large stationary sources. Many countries have been tightening air quality regulations to force significant emissions reductions of air pollutants such as sulphur dioxide, particulate matter, and mercury. Compliance with these regulations can also produce

co-benefits in terms of GHG reductions. These co-benefits and the channels through which they arise are examined, drawing on the experience of the European Union, the United States and Canada, as well as other regions. The results can be quite small or quite large, depending on factors that include the relative economics of coal- and gas-fired power generation and future expectations related to carbon control. The benefits of multi-pollutant strategies that take an integrated approach are underscored.

- China's air quality constraints: Implications for GHG mitigation in power and key industry sectors. China's "war on air pollution" agenda can drive ancillary reductions in carbon dioxide (CO_2) emissions and lead to the development of complementary air pollution and low-carbon policies. However, regional variation in pollution control measures and the design of industrial policies and measures in power and key industry sectors may limit GHG benefits (i.e. through geographic dislocation of emissions, methane or CO_2 leaks or increased CO_2 emissions intensity of pollution-reducing technologies). This is especially true if competing lower-carbon technology options do not provide for security of supply, or if air quality measures or monitoring and enforcement do not take a comprehensive accounting of environmental impacts. Overall, air pollution controls can lead to meaningful reductions in GHG emissions, provided they are structured to achieve these dual objectives.

- The regulatory approach to climate policy in the United States. To advance its climate change goals, the US government is targeting GHG emission reductions through a sectoral approach, using a regulatory framework normally reserved for the control of conventional air pollutants. The cornerstone of this approach is the application of federal carbon pollution standards to the electric power sector. Though the use of regulatory standards is a notable expansion of the climate policy toolkit beyond the market mechanisms that dominated much of the previous policy debate, they have been designed with some degree of market flexibility in mind. These GHG-targeted regulations also have important implications for air quality and public health co-benefits.

Source: https://www.iea.org/Textbase/npsum/EECC2014sum.pdf

Words and Phrases

intervention [ˌɪntə'venʃn] *n.* 干预；介入；调停；妨碍
[例句] He favoured a middle course between free enterprise and state intervention.
resurgence [rɪ'sɜːdʒəns] *n.* 回潮；再起；复苏，复活
[例句] Drugs traffickers are behind the resurgence of violence.
complementary [ˌkɒmplɪ'mentri] *adj.* 互补的；补充的
[例句] To improve the quality of life through work, two complementary strategies are necessary.
renewable [rɪ'njuːəbl] *adj.* 可再生的；可继续的
[例句] A formal contract is signed which is renewable annually.
nexus ['neksəs] *n.* (事物、思想等之间的)联结，联系
[例句] The Prayer Book has provided a flexible enough nexus of beliefs to hold together the different church parties.
compensate ['kɒmpenseɪt] *vt.* 抵消；补偿，赔偿；报酬；*vi.* 补偿，弥补
[例句] The company agreed to keep up high levels of output in order to compensate for supplies lost.
decarbonization [diːˌkɑːbənaɪ'zeɪʃən] *n.* 脱碳(作用)；除碳(法)
[例句] In this paper, decarbonization of silicon carbide platelets by using a novelrolling oxidation furnace was studied.
mandate ['mændeɪt] *n.* 授权；命令；委任；任期；*vt.* 托管；批准；*vi.* 强制执行；委托办理
[例句] The President and his supporters are almost certain to read this vote as a mandate for continued economic reform.

energy security 能源安全；能源保障；能源安全性
climate change 气候变化
energy system 能源系统
GHG emission 温室气体排放
short-term impacts 短期影响
air quality 空气质量
sulphur dioxide 二氧化硫
particulate matter 颗粒物
carbon dioxide 二氧化碳

Questions

1. What should we do to meet the challenge of climate change?
2. What lessons can be drawn from recent ETS experiences?
3. Why do many countries use energy sector goals and metrics to support GHG

emissions reductions?

Part C   Dialogue

Dialogue 1

Professor: We're really just now beginning to understand how quickly drastic climate change can take place. We can see past occurrences of climate change that took place over just a few hundred years. Take uh… the Sahara Desert in Northern Africa. The Sahara was really different 6,000 years ago. I mean, you wouldn't call it a tropical paradise or anything, uh…or maybe you would if you think about how today in some parts of the Sahara it…it only rains about once a century. Um… but basically, you had granary and you had water. And what I find particularly interesting and amazing really, what really indicates how desert-like the Sahara was thousands of years ago, was something painted on the rock, pre-historic art, hippopotamuses, as you know hippos need a lot of water and hence? Hence what?

Student: They need to live near a large source of water year round.

Professor: That's right.

Student: But how is that proved that the Sahara used to be a lot wetter? I mean the people who painted those hippos, well, couldn't they have seen them on their travels?

Professor: Okay, in principal they could, Karl. But the rock paintings aren't the only evidence. Beneath the Sahara are huge aquifers, basically a sea of fresh water, that's perhaps a million years old filtered through rock layers. And…er…and then there is fossilized pollen, from low shrubs and grasses that once grew in the Sahara. In fact these plants still grow, er…but hundreds of miles away, in more vegetated areas. Anyway, it's this fossilized pollen along with the aquifers and the rock paintings, these three things are all evidence that the Sahara was once much greener than it is today, that there were hippos and probably elephants and giraffes and so on. So what happened? How did it happen? Well, now, we're so used to hearing about how human activities are affecting the climate, right? But that takes the focus away from the natural variations in the earth climate, like the Ice Age, right? The planet was practically covered in ice just a few thousand years ago. Now as far as the Sahara goes, there is some recent literature that points to the migration of the monsoon in that area.

Students: Huh?????

Professor: What do I mean? Okay, a monsoon is a seasonal wind that can bring in a large amount of rainfall. Now if the monsoon migrates, well, that means that the rains move to another area, right? So what caused the monsoon to migrate? Well, the answer is: the dynamics of earth's motions, the same thing that caused the Ice Age by the way. The earth's not always the same distance from the sun, and it's not always tilting toward the sun at the same angle. There are slight variations in these two perimeters. They're gradual variations but their effects can be pretty abrupt. And can cause the climate to change in just a few hundred years.

Student: That's abrupt?

Professor: Well, yeah, considering that other climate shifts take thousands of years, this one is pretty abrupt. So these changes in the planet's motions, they called it "the climate change", but it was also compounded. What the Sahara experienced was um…a sort of "runaway drying effect". As I said the monsoon migrated itself, so there was less rain in the Sahara. The land started to get drier, which in turn caused huge decrease in the amount of vegetation, because vegetation doesn't grow as well in dry soil, right? And then, less vegetation means the soil can't hold water as well, the soil loses its ability to retain water when it does rain. So then you have less moisture to help clouds form, nothing to evaporate for cloud formation. And then the cycle continues, less rain, drier soil, less vegetation, fewer clouds, less rain etc.

Student: But, what about the people who made the rock paintings?

Professor: Good question. No one really knows. But there might be some connections to ancient Egypt. At about the same time that the Sahara was becoming a desert…

Student: Uh-huh

Professor: 5,000 years ago, Egypt really began to flourish out in the Nile River valley. And that's not that far away. So it's only logical to hypothesize that a lot of these people migrated to the Nile valley when they realized that this was more than a temporary drought. And some people take this a step further. And that's okay, that's science and they hypothesize that this migration actually provided an important impetus in the development of ancient Egypt. Well, we'll stay tuned on that.

Source: http://www.kekenet.com/menu/201307/250156.Shtml

Dialogue 2

Lecturer: So since we're around the topic of global climate change and its effects, in Alaska, in the northern Arctic part of Alaska, over the last thirty years or so, temperature has increased about half a degree Celsius per decade, and scientists have noticed that there've been changes in surface vegetation during this time. Shrubs are increasing in the "tundra". Tundra is flat land with very little vegetation. Just a few species of plants grow there because the temperature is very cold, and there's not much precipitation. And because of the cold temperatures, the tundra has two layers: top layer, which is called the active layer, is frozen in the winter and spring, but thaws in the summer. Beneath this active layer is the second layer called "permafrost", which is frozen all year around, and is impermeable to water.

Female Student: So because of the permafrost, none of the plants that grow there can have deep roots, can they?

Lecturer: No, and that's one of the reasons that shrubs survive in the Arctic. Shrubs are little bushes. They're not tall and being low in the ground protects them from the cold and wind. And their roots don't grow very deep, so the permafrost doesn't interfere with their growth. OK? Now since the temperatures have been increasing in Arctic Alaska, the growth of shrubs has increased. And this is presented to climate scientists with a puzzle...

Male Student: I'm sorry, when you say the growth of shrubs has increased, do you mean the shrubs are bigger, or that there are more shrubs?

Lecturer: Good question! And the answer is both. The size of the shrubs has increased and shrub cover has spread to what was previously shrub-free tundra. Ok, so what's the puzzle? Warmer temperatures should lead to increased vegetation growth, right? Well, the connections are not so simple. The temperature increase has occurred during the winter and spring, not during the summer. But the increase in shrubs has occurred in the summer. So how can increase temperatures in the winter and spring result in increased shrub growth in the summer? Well, it may be biological processes that occur in the soil in the winter, that cause increased shrub growth in the summer, and here's how: there are "microbes", microscopic organisms that live in the soil. These microbes enable the soil to have more nitrogen, which plants need to live and they remain quite active during the winter. There're two reasons for this: first, they live in the active layer, which, remember, contains water that doesn't penetrate the permafrost. Second, most of the precipitation in the Arctic is in the form of snow. And the snow, which blankets the ground in the winter, actually has an insulating effect on the soil beneath it. And it allows the temperature of the soil to remain warm enough for microbes to remain active. So there's been increase in nutrient production in the winter. And that's what's responsible for the growth of shrubs in the summer and their spread to new areas of the tundra. Areas with more new nutrients are the areas with the largest increase in shrubs.

Female student: But, what about run-off in the spring, when the snow finally melts? Won't the nutrients get washed away? Spring thaw always washes away soil, doesn't it?

Lecturer: Well, much of the soil is usually still frozen during peak run-off. And the nutrients are deep down in the active layer anyway, not high up near the surface, which is the part of the active layer most affected by run-off. But as I was about to say, there's more to the story. The tundra is windy, and the snow is blown across the tundra, it's caught by shrubs. And deep snow drifts often form around shrubs. And we've already mentioned the insulating effect of snow. So that extra warmth means even more microbial activities, which mean even more food for the shrubs, which mean even more shrubs and more snow around etc. It's a circle, a loop. And because of this loop, which is promoted by warmer temperatures in winter and spring, well, it looks like the tundra may be turning into shrub land.

Female student: But will it be long term? I mean maybe the shrubs will be abundant for a few years, and then it'll change back to tundra.

Lecturer: Well, shrub expansion has occurred in other environments, like semiarid grassland, and tall grass prairies. And shrub expansion in these environments does seem to persist, almost to the point of causing a shift. Once is established, shrub land thrives, particularly in the Arctic, because Arctic shrubs are good at taking advantage of increased nutrients in the soil, better than other Arctic plants.

Source: http://www.lasedu.com/laseduhtml/toefl/listening/7276.html

Part D　Supplementary Vocabulary

Energy and Climate Change

alternator 发电机
anthracite 无烟煤
automotive carbon emission 汽车碳排放
bioenergy 生物能
biomass energy 生物质能
bitumite 烟煤
carbon intensity 碳强度
carbon-abatement project 碳减排项目
carbon-free energy 无碳能源
carbon-intensive energy source 高碳能源
carbon-intensive industry 高碳行业
clean energy 清洁能源
climate disaster 气候灾难
common but differentiated responsibility 共同但有区别的责任
consumption per head 人均消费量
cumulative carbon emission 累计碳排放
current emission 当前排放
cut emission from electricity production 减少发电的排放
decarbonize the economy 经济去碳化
diesel 柴油
emission perunit of GDP 单位GDP的排放量
emissions-cutting requirement 减排

要求
energy consumption 能源耗量
energy demand 能源需求
energy development 能源开发
energy policy 能源政策
energy source 能源
faeces 生物粪便
forest carbon sink 森林碳汇
gasoline 汽油
geothermal energy 地热能
global carbon emission 全球碳排放
global warming 全球变暖
historic emission 历史排放
hydroelectric 水电的
hydroenergy 水能
international climate change negotiation 国际气候变化谈判
low-carbon technology 低碳技术
major emitter 排放大国
mazout 重油
melting sea ice 海冰融化
mineral reserve 矿产资源
mitigate and adapt to climate change 减缓和适应气候变化
natural gas 天然气
nuclear energy 核能
ocean energy 海洋能
peat land 泥炭地
petroleum 石油

power station 发电厂
primary energy 一次能源
radioactive 放射性的
rational utilization 合理利用
reserve 储量
resource 资源
secondhand energy 二次能源
set a carbon intensity target 确定碳强度的指标
solar energy 太阳能
steam 蒸汽
the Bali Roadmap 巴厘路线图
the biggest carbon emitter per person 人均碳排放最多的国家
the Clean Development Mechanism (CDM) 清洁发展机制
the Copenhagen Summit on Climate Change 哥本哈根气候变化峰会
the world's major carbon-emitting nation 世界主要碳排放国
thermal pollution 热污染
tidal energy 潮汐能
use ratio 利用率
vulnerability assessment 脆弱性评估
water resource 水资源
wind energy 风能
zero-emission power source 零排放电力来源

Part E Supplementary Knowledge

有机化合物命名法

1. 有机化合物的官能团分类

物质的性质与物质的结构密切相关。体现有机化合物主要结构特征的是其分子中的官能团(functional group)。官能团决定着有机化合物的性质，所以按照官能团来对有机化合物进行分类是有机化学常用的一种分类方法。根据不同的官能团，有机化合物主要可分为：烃(hydrocarbon)、醇(alcohol)、醚(ether)、醛(aldehyde)、酮(ketone)、羧酸(carboxylic acid)、酯(ester)、胺(amine)、酰胺(amide)、氨基酸(amino acid)和腈(nitrile)等。现在国际上通常用的是国际纯粹与应用化学联合会系统命名法，简称 IUPAC 系统命名法。

2. 烃的命名

1) 烷烃(alkane)：有关词头 + -ane

例：
- CH_4 methane
- CH_3CH_3 ethane
- $CH_3CH_2CH_3$ propane
- $CH_3CH_2CH_2CH_3$ butane

从戊烷(pentane)开始，烷烃的命名根据其含碳数由希腊数字派生(表 1)。如果希腊数字末尾带字母-a，命名对应的烷烃时，直接在其后加-ne。把烷烃名中词尾-ane 用-yl 取代，就形成了对烷基的命名(表 2)。

表 1 命名有机物的数字词头

n	前缀 Prefixes	N	前缀 Prefixes	n	前缀 Prefixes
1	mon-	12	dodeca-	23	tricosa-
2	di-	13	trideca-	30	triaconta-
3	tri-	14	tetradeca-	31	hentriaconta-
4	tetra-	15	pentadeca-	40	tetraconta-
5	penta-	16	hexadeca-	50	pentaconta-
6	hexa-	17	heptadeca-	60	hexaconta-
7	hepta-	18	octadeca-	70	heptaconta-
8	octa-	19	nonadeca-	80	octaconta-
9	nona-	20	eicosa-	90	nonaconta-
10	deca-	21	heneicosa-	100	hecta-
11	undeca-	22	docosa-		

表 2 一些烷基的命名

化学式 Formula	名称 Name	化学式 Formula	名称 Name
CH_3-	methyl	$(CH_3)_2C-$	tert-butyl
CH_3CH_2-	ethyl	$CH_3(CH_2)_3CH_2-$	n-pentyl
$CH_3CH_2CH_2-$	n-propyl	$CH_3(CH_2)_5CH_2-$	n-heptyl

续表

化学式 Formula	名称 Name	化学式 Formula	名称 Name
$CH_3(CH_2)_7CH_2$-	n-nonyl	$CH_3(CH_2)_6CH_2$-	n-butyl
$CH_3CH(CH_3)$-	isoproply	$CH_3(CH_2)_6CH_2$-	n-hexyl
$(CH_3)_2CHCH_3$-	isobutyl	$CH_3(CH_2)_7CH_2$-	n-octyl
$CH_3CH_2CH_2CH_2$-	sec-butyl	$CH_3(CH_2)_8CH_2$-	n-decyl

"n"即"normal",相当于中文的"正",表示烷基中无支链;"iso"相当于中文的"异",通常指烷基的一端有$(CH_3)_2CH$-结构;"sec""tert"即"secondary""tertiary",相当于中文的"仲"和"叔",分别表示该基团以其中的"仲"和"叔"碳原子和别的基团相连接。

2) 烯烃和炔烃

烯烃和炔烃的命名可由对应的烷烃名而衍生,只要把烷烃名后的-ane 用-ene 或-yne 取代。当分子中有多个相同的基团或官能团时,分别用 di(二)、tri(三)、tetra(四)等说明。

3) 芳香烃

芳香烃因其分子中含有不饱和碳碳双键,其英文名都以-ene 结尾。如苯,benzene;萘,naphthalene;蒽,anthracene;菲,phenanthrene 等。

芳香烃衍生物的命名:若将芳香烃作为母体,则在芳香烃名(如 benzene,苯)前加上取代基的位置、数目和名字即成。若芳香烃作为取代基,则称为芳香烃基(如 phenyl 或 ph-,苯基)。其中,苯环上多个取代基的相对位置除了可用数字编号来说明外,还可用 ortho-或 o-(邻)、meta-或 m-(间)、para-或 p-(对)来表示。例:

m-(或1,3-)dimethyl benzene

3. 醇和醚的命名

对醇而言,英文中普通名和 IUPAC 系统名都是常用的。普通名通常是在"alcohol(醇)"字前加上连在—OH 上的有机基团名即成,IUPAC 名只要把对应的母体烷烃名词尾-ane 中的-e 用-ol 取代即可(表3)。

表3 一些醇的命名

种类 Class	化学式 Formula	国际化联名称 IUPAC Name	普通名称 Common Name
primary	CH_3OH	methanol	methyl alcohol
primary	CH_3CH_2OH	etanol	erhyl alcohol

续表

种类 Class	化学式 Formula	国际化联名称 IUPAC Name	普通名称 Common Name
primary	$CH_3(CH_2)_3CH_2OH$	1-pentanol	(n-amyl or n-pentyl alcohol)
primary	$CH_3(CH_2)_6CH_2OH$	1-octanol	n-octyl alcohol
primary	$(CH_3)_2CHCH_2OH$	2-methyl-1-propanol	isobutyl alcohol
secondary	$(CH_3)_2CHOH$	2-propanol	isopropyl alcohol
tertiary	$C(CH_3)_3OH$	2-methyl-2-propanol	tert-butyl alcohol

命名多元醇时，是在词尾-ol 前加上 di-、tri-等来表示羟基 OH 的数目，但母体烷烃名词尾-ane 中的 e 要保留（因为 di、tri 中的 d 和 t 是辅音，而 ol 中的 o 是元音）。例：

$$\begin{array}{cc} CH_2-CH_2 \\ | \quad | \\ OH \quad OH \\ \text{1,2-ethanediol} \end{array} \qquad \begin{array}{ccc} CH_2-CH-CH_2 \\ | \quad | \quad | \\ OH \quad OH \quad OH \\ \text{1,2,3-propanetriol} \end{array}$$

简单的醚习惯上几乎无例外地由普通法命名，在"ether（醚）"前加上 2 个连接到氧原子上的有机基团即成，例：

$CH_3—O—CH_2CH_3$　　　　　　　　　　ethyl methyl ether

$CH_3—O—CH_3$　　　　　　　　　　　　methyl ether or dimethyl ether

较复杂醚的命名多采用 IUPAC 法，把烷氧基（RO—）作为取代基看待，如：

$CH_3OCH_2CH_2OCH_3$　　　　　　　　　1,2-dimethoxyethane

$CH_3CH_2OCH_2CH_2CH_2CH(OH)CH_3$　　5-ethoxy-2-pentanol

4. 醛和酮的命名

醛和酮是两类具有相似结构的有机化合物，在 IUPAC 系统命名中，醛、酮的命名只需把母体烃名词尾中的-e 分别用-al 或-one 取代即成（多元醛、酮，则保留烃的词尾-e，再在-al，-one 前加 di-、tri-等表示醛基和羰基的数目）。例：CH_3CHO 为 ethanal。

5. 羧酸和酯的命名

在 IUPAC 系统命名中，羧酸的命名是把母体烃名词尾中的-e 用-oic 取代（命名多元羧酸时则保留烃的词尾-e），并在其后加"acid（酸）"。例：

HCOOH　　　　　　　　methanoic acid

HOOC—CH_2—COOH　　propanedioic acid

简单的醛、酮和羧酸通常用普通法命名。把羧酸普通名中的词尾-ic acid 改成 aldehyde，就变成了对应醛的普通名。而酮的普通名则是在"ketone"前加上 2 个

烃基名(按第一个英文字母排序),如下列羧酸、醛和酮的普通名分别是:

$$
\begin{array}{ll}
\text{HCOOH} & \text{formic acid} \\
\text{CH}_3\text{COOH} & \text{acetic acid} \\
\text{HCHO} & \text{formaldehyde} \\
\text{CH}_3\text{CHO} & \text{acetaldehyde} \\
\text{CH}_3\overset{\text{O}}{\underset{\|}{\text{C}}}\text{CH}_2\text{CH}_3 & \text{ethyl methylketone}
\end{array}
$$

酯的命名如同羧酸盐,带词尾-ate,它的最前面是醇的母体烃基名,中间是去掉-ic acid 词尾的羧酸名,最后是表示酯类的词尾-ate,例:

$$\text{CH}_3-\underset{\underset{\text{O}}{\|}}{\text{C}}-\text{O}-\text{CH}_3 \quad \text{methyl ethanoate (IUPAC name) or methyl acetate (common name)}$$

6. 胺、酰胺、氨基酸和腈的命名

简单的胺,命名时通常把胺基作为官能团,英文名是在 amine 前加上烃基名(烃基名按第一个字母序写出)。二胺、三胺则在-diamine、-triamine 前加上母烃的名(注:不是烃基的名)。若有多种烃基连在氮原子上,命名时则选取含氮的最长链作为母体,氮上的其他烃基作为取代基,并用 N 定其位,例:

$$
\begin{array}{ll}
(\text{CH}_3\text{CH}_2)_2\text{NH} & \text{diethylamine} \\
\text{H}_2\text{NCH}_2\text{CH}_2\text{CH}_2\text{NH}_2 & \text{1,3-propanediamine} \\
\text{H}_3\text{C}-\overset{\text{CH}_3}{\underset{|}{\text{CH}}}\text{NHCH}_3 & \text{N-methyl-2-propylamine}
\end{array}
$$

氮原子直接与酰基相连,称为酰胺,可看作是羧酸的一种衍生物,英文名是将对应的羧酸词尾中的-oic acid 改为-amide。例:

$$\text{CH}_3\overset{\text{O}}{\underset{\|}{\text{C}}}-\text{NH}_2 \quad \text{ethanamide}$$

腈可以水解成羧酸,因此也可以把腈看作羧酸的衍生物。英文名是将对应的羧酸词尾中的-oic acid 改为-onitrile。例:

$$
\begin{array}{ll}
\text{CH}_3\text{CN} & \text{ethanonitrile} \\
\text{CH}_3\text{CH}_2\text{CN} & \text{propanonitrile}
\end{array}
$$

氨基酸,是羧酸的氨基(amino)取代物,英文名是在对应的羧酸名前加上表示氨基及其位置的词头。例:

$$\text{CH}_3-\underset{\underset{\text{NH}_2}{|}}{\text{CH}}-\text{CO}_2\text{H} \quad \text{2-aminopropanoic acid}$$

有机化合物的常见基团都有作词尾、词头两种可能，作词尾时相当于母体官能团，作词头时相当于取代基，它们的英文表达形式是不同的，这是英文命名时尤其要注意的。例如—SO_3H，—CN，—OH，—NH_2，作词头时分别是 sulfo-（磺酸基），cyano-（氰基），hydroxy-（羟基），amino（氨基）；作词尾时分别是-sulfonic acid（磺酸），carbonitrile（腈），-ol（醇）和-amine（胺）。

Reference: 张文广，王祖浩. 有机化合物的英文命名法[J]. 化学教育，2006, 27(11):38-40.

Unit 4　Air Pollution

Part A　Intensive Reading

Air Pollution

The atmosphere is a complex natural gaseous system that is essential to support life on planet earth. Air pollution is the introduction of particulates, biological molecules, and many harmful substances into earth's atmosphere, causing diseases, allergies, death to humans, damage to other living organisms such as animals and food crops, or the natural or built environment. Air pollution may come from anthropogenic or natural sources.

Pollutants

An air pollutant is a substance in the air that can have adverse effects on humans and the ecosystem (Figure 1). The substance can be solid particles, liquid droplets, or gases. A pollutant can be of natural origin or man-made. Pollutants are classified as primary or secondary. Primary pollutants are usually produced from a process, such as ash from a volcanic eruption. Other examples include carbon monoxide gas from motor vehicle exhaust, or the sulfur dioxide released from factories. Secondary pollutants are not emitted directly. Rather, they form in the air when primary pollutants react or interact. Ground level ozone is a prominent example of a secondary pollutant. Some pollutants may be both primary and secondary: they are both emitted directly and formed from other primary pollutants.

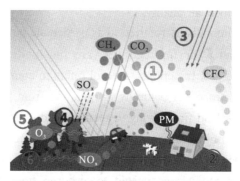

Figure 1　Causes and effects of air pollution
① greenhouse effect；② particulate contamination；③ increased UV radiation；④ acid rain；
⑤ increased ground level ozone concentration；⑥ increased levels of nitrogen oxides

Major primary pollutants produced by human activity include:

- Carbon dioxide (CO_2)—Because of its role as a greenhouse gas it has been described as "the leading pollutant" and "the worst climate pollution". Carbon dioxide is a natural component of the atmosphere, essential for plant life and given off by the human respiratory system.
- Sulfur oxides (SO_x), particularly sulfur dioxide, a chemical compound with the formula SO_2. SO_2 is produced by volcanoes and in various industrial processes. Coal and petroleum often contain sulfur compounds, and their combustion generates sulfur dioxide. Further oxidation of SO_2, usually in the presence of a catalyst such as NO_2, forms H_2SO_4, and thus acid rain.
- Nitrogen oxides (NO_x)—Nitrogen oxides, particularly nitrogen dioxide, are expelled from high temperature combustion, and are also produced during thunderstorms by electric discharge. Nitrogen dioxide is a chemical compound with the formula NO_2. It is one of several nitrogen oxides. One of the most prominent air pollutants, this reddish-brown toxic gas has a characteristic sharp, biting odor.
- Carbon monoxide (CO)—CO is a colorless, odorless, toxic yet non-irritating gas. It is a product of incomplete combustion of fuel such as natural gas, coal or wood. Vehicular exhaust is a major source of carbon monoxide.
- Volatile organic compounds (VOCs)—VOCs are a well-known outdoor air pollutant. They are categorized as either methane (CH_4) or non-methane (NMVOCs). Methane is an extremely efficient greenhouse gas which contributes to enhanced global warming. Other hydrocarbon VOCs are also significant greenhouse gases because of their role in creating ozone and prolonging the life of methane in the atmosphere. This effect varies depending on local air quality.
- Particulates, alternatively referred to as particulate matter (PM), atmospheric particulate matter, or fine particles, are tiny particles of solid or liquid suspended in a gas. In contrast, aerosol refers to combined particles and gas. Some particulates occur naturally, originating from volcanoes, dust storms, forest and grassland fires, living vegetation, and sea spray. Human activities, such as the burning of fossil fuels in vehicles, power plants and various industrial processes also generate significant amounts of aerosols.
- Persistent free radicals connected to airborne fine particles are linked to cardiopulmonary disease.

- Toxic metals, such as lead and mercury, especially their compounds.
- Chlorofluorocarbons (CFCs)—CFCs are harmful to the ozone layer; emitted from products are currently banned from use. These are gases which are released from air conditioners, refrigerators, aerosol sprays, etc. On release into the air, CFCs rise to the stratosphere. Here they come in contact with other gases and damage the ozone layer. This allows harmful ultraviolet rays to reach the earth's surface.
- Ammonia (NH_3)—Ammonia is a compound with the formula NH_3 and is emitted from agricultural processes. It is normally encountered as a gas with a characteristic pungent odor. Ammonia contributes significantly to the nutritional needs of terrestrial organisms by serving as a precursor to foodstuffs and fertilizers. Ammonia, either directly or indirectly, is also a building block for the synthesis of many pharmaceuticals.
- Odours—such as from garbage, sewage, and industrial processes.
- Radioactive pollutants—Radioactive pollutants are produced by nuclear explosions, nuclear events, war explosives, and natural processes such as the radioactive decay of radon.

Secondary pollutants include:
- Particulates created from gaseous primary pollutants and compounds in photochemical smog. Smog is a kind of air pollution. Classic smog results from large amounts of coal burning in an area caused by a mixture of smoke and sulfur dioxide. Modern smog does not usually come from coal but from vehicular and industrial emissions that are acted on in the atmosphere by ultraviolet light from the sun to form secondary pollutants that also combine with the primary emissions to form photochemical smog.
- Ground level ozone (O_3) formed from NO_x and VOCs. Ozone (O_3) is a key constituent of the troposphere. It is also an important constituent of certain regions of the stratosphere commonly known as the Ozone layer. Photochemical and chemical reactions involving it drive many of the chemical processes that occur in the atmosphere by day and by night. At abnormally high concentrations brought about by human activities (largely the combustion of fossil fueles), it is a pollutant, and a constituent of smog.
- Peroxyacetyl nitrate (PAN)—similarly formed from NO_x and VOCs.

Minor air pollutants include:
- A large number of minor hazardous air pollutants. Some of these are regulated

in USA under the Clean Air Act and in Europe under the Air Framework Directive.
- A variety of persistent organic pollutants, which can attach to particulates.

Persistent organic pollutants (POPs) are organic compounds that are resistant to environmental degradation through chemical, biological, and photolytic processes. They have been observed to persist in the environment, to be capable of long-range transport, bioaccumulate in human and animal tissue, biomagnify in food chains, and to have potentially significant impacts on human health and the environment.

There are 12 compounds in the list of persistent organic pollutants. Dioxins and furans are two of them and intentionally created by combustion of organics, like open burning of plastics. These compounds are also endocrine disruptors and can mutate the human genes.

Sources

There are various locations, activities or factors which are responsible for releasing pollutants into the atmosphere. These sources can be classified into two major categories.

Anthropogenic (man-made) sources:

These are mostly related to the burning of multiple types of fuel.
- Stationary sources include smoke stacks of power plants, manufacturing facilities (factories) and waste incinerators, as well as furnaces and other types of fuel-burning heating devices. In developing and poor countries, traditional biomass burning is the major source of air pollutants; traditional biomass includes wood, crop waste and dung.
- Mobile sources include motor vehicles, marine vessels, and aircraft.
- Controlled burn practices in agriculture and forest management. Controlled or prescribed burning is a technique sometimes used in forest management, farming, prairie restoration or greenhouse gas abatement. Fire is a natural part of both forest and grassland ecology and controlled fire can be a tool for foresters. Controlled burning stimulates the germination of some desirable forest trees, thus renewing the forest.
- Fumes from paint, hair spray, varnish, aerosol sprays and other solvents.
- Waste deposition in landfills generates methane. Methane is highly flammable and may form explosive mixtures with air. Methane is also an asphyxiant and may displace oxygen in an enclosed space. Asphyxia or suffocation may result if the oxygen concentration is reduced to below 19.5% by displacement.

- Military resources, such as nuclear weapons, toxic gases, germ warfare and rocketry.

Natural sources:
- Dust from natural sources, usually large areas of land with little or no vegetation.
- Methane, emitted by the digestion of food by animals, for example cattle.
- Radon gas from radioactive decay within the earth's crust. Radon is a colorless, odorless, naturally occurring, radioactive noble gas that is formed from the decay of radium. It is considered to be a health hazard. Radon gas from natural sources can accumulate in buildings, especially in confined areas such as the basement and it is the second most frequent cause of lung cancer, after cigarette smoking.
- Smoke and carbon monoxide from wildfires.
- Vegetation, in some regions, emits environmentally significant amounts of VOCs on warmer days. These VOCs react with primary anthropogenic pollutants—specifically, NO_x, SO_2, and anthropogenic organic carbon compounds—to produce a seasonal haze of secondary pollutants. Black gum, poplar, oak and willow are some examples of vegetation that can produce abundant VOCs. The VOC production from these species result in ozone levels up to eight times higher than the low-impact tree species.
- Volcanic activity, which produces sulfur, chlorine, and ash particulates.

Source: https://en.wikipedia.org/wiki/Air_pollution

Words and Phrases

particulate [pɑ:'tɪkjələt]　*adj.* 微粒的，颗粒的；*n.* 微粒，粒子
anthropogenic [ˌænθrəpə'dʒnɪk] *adj.* 人类起源的；人为的
exhaust [ɪg'zɔ:st]　*vt.* 排出　*vi.* 排气；*n.* (排出的)废气，排出，排气装置
petroleum [pə'trəʊliəm]　*n.* 石油
combustion [kəm'bʌstʃən]　*n.* 燃烧，燃烧过程
oxidation [ˌɒksɪ'deɪʃn]　*n.* 氧化
catalyst ['kætəlɪst]　*n.* [化]催化剂；促使变化的人，引发变化的因素
expel [ɪk'spel]　*vt.* 排出(气体等)
reddish ['redɪʃ]　*adj.* 略带红色的，微红的
aerosol ['eərəsɒl]　*n.* [化]气溶胶，喷雾剂
cardiopulmonary [ˌkɑ:dɪəʊ'pʌlmən(ə)rɪ] *adj.* [医]心肺的
stratosphere ['strætəsfɪə(r)]　*n.* [气]平流层，同温层
ammonia [ə'məʊniə]　*n.* 氨，氨水，氨气

pungent ['pʌndʒənt]　*adj*. 味道强烈的，刺激性的
synthesis ['sɪnθəsɪs]　*n*. [化]合成，综合体
pharmaceutical [ˌfɑːməˈs(j)uːtɪkl]　*adj*. 制药的，配药的；*n*. 药物
sewage ['suːɪdʒ]　*n*. (下水道里的)污水，下水道，污水
radon ['reɪdɒn]　*n*. [化]氡(元素符号Rn)
troposphere ['trɒpəsfɪə(r)]　*n*. 对流层
photolytic [fəʊtə'lɪtɪk]　*adj*. 光解的
biomagnify ['baɪɒmægnɪfaɪ]　*vi*. (有毒化学物质)产生生物放大作用
dioxin [daɪ'ɒksɪn]　*n*. 二噁英
furan ['fjʊərən]　*n*. 呋喃
endocrine ['endəʊkrɪn]　*adj*. 内分泌腺的，内分泌的
incinerator [ɪn'sɪnəreɪtə(r)]　*n*. (废物的)焚化炉
furnace ['fɜːnɪs]　*n*. 熔炉，火炉，极热的地方，严峻的考验或磨难
prairie ['preəri]　*n*. 北美草原；新大陆北部草原(美国北部和加拿大)
flammable ['flæməbl]　*adj*. 易燃的，可燃的
asphyxia [æs'fɪksɪə]　*n*. 血液中缺氧，窒息
asphyxiant ['əsfɪksjənt]　*adj*. 导致窒息的；*n*. 引起窒息的事物

suffocation [ˌsʌfə'keɪʃn]　*n*. 窒息而死
radium ['reɪdiəm]　*n*. [化]镭(88号元素符号Ra)
hazard ['hæzəd]　*n*. 危险，危害
allergy ['ælədʒi]　*n*. 过敏反应；变态反应
solid particle ['sɒlɪd 'pɑːtɪkl]　固体微粒，固体粒子
volcanic eruption [vɒl'kænɪk ɪ'rʌpʃn]　火山爆发
carbon monoxide [kɑːbən mə'nɒksaɪd]　一氧化碳
sulfur dioxide　二氧化硫
incomplete combustion　不完全燃烧
volatile organic compounds（VOCs）挥发性有机化合物
grassland fire　草原火灾
dust storm　沙尘暴
free radical ['rædɪkl]　自由基，游离基
ultraviolet rays　[医]紫外线
photochemical smog　光化学烟雾
peroxyacetyl nitrate（PAN）[pərɒksɪ'æsɪtɪl 'naɪtreɪt]　硝酸过氧化乙酰
persistent organic pollutants（POPs）持久性有机污染物
stationary source　固定污染源
noble gas ['nəʊbl]　惰性气体

Questions

1. What is air pollution?

2. What are the major pollutants produced by anthropogenic activities (both primary and secondary)?

3. List some examples of sources, where air pollutants are produced.

4. What are air pollution emission factors?

5. How many kinds of POPs in the air are mentioned in the text?

Sentence-making

1. particulates, anthropogenic, natural
2. carbon monoxide, sulfur dioxide, combustion
3. catalyst, NO_x, vehicle
4. ultraviolet rays, free radical, photochemical smog
5. pungent, dioxin, furan

Part B   Extensive Reading

Industrial Air Pollution Leaves Magnetic Waste in the Brain

If you live in an urban environment, chances are you've got nanomagnets on the brain—literally. New research suggests that most magnetite found in the human brain, a magnetic iron oxide compound, comes from industrial air pollution. And because unusually high concentrations of magnetite are found in the brains of people with Alzheimer's disease, the findings raise the specter of an alarming new environmental risk factor for this and other neurodegenerative diseases. Still, other scientists caution that the link remains speculative.

For decades, scientists have known the brain harbors magnetic particles, but most assumed that they derived naturally from the iron used in normal brain function. About 25 years ago, geophysicist Joe Kirschvink at California Institute of Technology in Pasadena detected biologically formed magnetite particles in human brains, lending evidence to their natural origin.

The problem with magnetite is that it's toxic. It causes oxidative stress, disrupting normal cellular function and contributing to the creation of destructive free radicals—unstable molecules that can damage other important molecules. Previous work has also shown a correlation between high amounts of brain magnetite and Alzheimer's disease, and recent studies suggest it increases the toxicity of the disease's hallmark β amyloid plaques, clumps of protein that can interfere with cell signaling. Nothing definitively links magnetite to Alzheimer's, but the kinds of cellular damage it can cause are consistent with what's seen in the diseases.

Physicist Barbara Maher, co-director of the Centre for Environmental Magnetism and Paleomagnetism at Lancaster University in the United Kingdom, wondered whether all the magnetite found in the brain could be traced to biological processes. As an expert in environmental magnetic particles, she knew that many, including magnetite, are prevalent in air pollution let off by power plant smokestacks.

"The paradigm until now has been that magnetite just forms naturally in the brain," Maher says. "Given how prolific magnetite particles are in the atmosphere, I wondered if they had gained entry into the human brain."

So Maher and a team of U.K. and Mexican scientists looked at postmortem samples of brain matter from the frontal cortexes of 37 human brains. Most came from people who lived in Mexico City, and others came from former residents of Manchester, U.K. Using a variety of high-resolution imaging techniques, the researchers examined the characteristics of the brain samples' magnetite.

Biological magnetite usually forms in tetrahedral or octahedral shapes, but the vast majority of magnetite particles found in this study were instead round nanospheres. The researchers detected other unexpected metallic nanoparticles in the brain samples, too: platinum, nickel, and cobalt. None of these metals occurs naturally in the brain, suggesting an environmental origin.

For Maher, the magnetite's nanosphere shape, combined with the presence of the other metals, was a tell-tale sign. "They showed all the properties suggesting they formed in high temperatures," she says. Those temperatures vary with the fuel being burned, but they are much higher than those of the human body. "The nanospheres are combustion byproducts, like what's found in power station pollution." Frictional heating, like what happens to a car's brake pads, can also produce magnetite nanospheres, she adds.

The researchers found biologically formed magnetite, too, but for every one of those particles, they found at least 100 pollution-derived magnetite particles, they report today in the *Proceedings of the National Academy of Sciences*. That comes out to millions of magnetite particles per gram of brain matter.

At 150 nanometers or less in diameter, these magnetite nanoparticles are small enough to be inhaled through the nose and enter the brain through the olfactory bulb. Previous air quality studies in the United Kingdom and Mexico City have found that urban areas, especially along roadsides, have abundant airborne magnetite, providing plenty of opportunity for people to sniff these toxic nanoparticles into their brain.

Other environmental risk factors for Alzheimer's and other neurodegenerative diseases have been suggested, and it's not clear how magnetite fits into the overall risk picture, or whether there's an exposure threshold that's especially dangerous. Still, the results deserve special attention from epidemiologists and air quality policymakers. Policymakers have tried to account for this in their environmental regulations, but maybe those need to be revised.

Kirschvink, the first scientist to detect biologically derived magnetite in the brain, is convinced that the researchers have indeed found evidence that magnetite can be introduced through air pollution. The especially high levels of magnetite found in the study's brain samples are shocking, though not necessarily surprising given that they lived and worked in industrial environments with lots of air pollution.

Researchers found so much environmental magnetite in the brain. Such magnetite is more dangerous than biogenic versions of the particle. That nanoparticles of industrially generated magnetite are able to make their way into the brain tissues is disturbing.

Jennifer Pocock, a neurologist at University College London, says in a statement through the independent Science Media Centre that although she agrees the evidence points to pollution-borne magnetite in the brain, more work is needed before a solid connection to Alzheimer's can be drawn. "There needs to be a better study carried out to correlate the magnetite concentrations of patients who lived primarily in a city, for example, compared with patients living in relatively unpolluted area, and analyses of their Alzheimer's disease brain pathology correlated with magnetite concentrations."

Source: Price M. Industrial air pollution leaves magnetic waste in the brain[J]. Science, 2016.

Words and Phrases

magnetite ['mægnɪtaɪt]　*n.* [矿物]磁铁矿
[例句] When these surfaces came into contact with the magnetite, an oxygen atom could be transferred from quartz to magnetite, forming a new mineral, hematite.
alzheimer's ['æltshaɪməz]　*n.* 老年痴呆症
[例句] Studies have linked a wide range of health benefits to green tea including lowering the risk of certain cancers and heart disease as well as weight loss and protection against Alzheimer's.
neurodegenerative [ˌnjuːrəʊdɪ'dʒenərətɪv]　*adj.* 神经变性的
[例句] In the National Alzheimer's Disease Awareness Month, more reports will be published here to help readers understand the age-related neurodegenerative disorder and how to prevent it.
cellular ['seljələ(r)]　*adj.* 细胞的；由细胞组成的；(无线电话)蜂窝状的

[例句] We may all be part of some larger body of humanity, but our interactions mainly occur at the individual cellular level.

radical ['rædɪkl]   *n.* 自由基；激进分子；[数]根基

[例句] In chemistry, a radical (more precisely, a free radical) is an atom, molecule, or ion that has an unpaired valence electron.

paleomagnetism [ˌpeɪlɪəʊ'mægnɪtɪzəm]   *n.* 古地磁学

[例句] Tarduno is a professor in the departments of earth and environmental sciences and of physics and astronomy at the University of Rochester and is founder of its paleomagnetism laboratory.

tetrahedral ['tetrə'hedrəl]   *adj.* 四面体的；有四面的

[例句] According to the conventional view, liquid water has a similar, albeitless rigid, structure, in which extra molecules can pack into some of the open gaps in the tetrahedral arrangement.

octahedral [ˌɒktə'hiːdrəl]   *adj.* 八面体的；有八面的

[例句] During the reaction, the octahedral dissolution of palygorskite induced heavy metal ions coprecipitation forming mixed double hydroxid precipitation on the surface of palygorskite.

platinum ['plætɪnəm]   *n.* [化学]铂；白金

[例句] In a laboratory vault outside Paris is a small cylinder of platinum–iridium alloy that serves as the standard for all mass measurements worldwide.

nickel ['nɪkl]   *n.* 镍；镍币；五分镍币(美国和加拿大)

[例句] They take an atom of lead, accelerate it and have it collide with an atom of nickel.

cobalt ['kəʊbɔːlt]   *n.* [化学]钴；深蓝色；钴蓝

[例句] The cobalt and phosphate form a thin-film catalyst around the electrode that then use electrons from the electrode to split the oxygen from water.

threshold ['θreʃhəʊld]   *n.* 入口；门槛；开始；极限；临界值

[例句] "We have crossed an important threshold," he said.

epidemiologist [ˌepɪˌdiːmɪ'ɒlədʒɪst]   *n.* 流行病学家

[例句] Because no vaccine protects 100 percent of kids who get it, epidemiologists rely on "herd immunity" to make sure enough kids are well enough protected to keep a disease from spreading.

neurologist [njʊə'rɒlədʒɪst]   *n.* 神经病学家；神经科医师

[例句] Animals (not just people) likely have spiritual experiences, according to a prominent neurologist who has analyzed the processes of spiritual sensation for over three decades.

pathology [pəˈθɒlədʒi]   *n.* 病理学（复数 pathologies）；变态；反常
[例句] Post-mortems showed the pathology—signs of disease—in the brains of people with and without long educations were at similar levels.

magnetic iron oxide compound 磁性氧化铁化合物
neurodegenerative disease 神经退行性疾病
magnetic particle 磁性微粒
oxidative stress 氧化应激
amyloid plaque 淀粉样蛋白斑
clumps of protein 蛋白质团块
environmental magnetism 环境磁学
environmental magnetic particle 环境磁性粒子
prolific magnetite particle 丰富的磁铁矿微粒
frontal cortex 额叶皮质
high-resolution imaging technique 高分辨率成像技术
biological magnetite 生物磁铁矿
instead round nanosphere 相反的圆形纳米微球
metallic nanoparticle 金属纳米微粒
magnetite's nanosphere shape 磁铁矿的纳米微球形状
frictional heating 摩擦加热
toxic nanoparticle 有毒的纳米微粒

Questions

1. Where does the magnetite in the human brain comes from?
2. Where are the hazards of magnetite in the brain to human health?
3. What are the characteristics of the brain samples' magnetite? And why they have these characteristics?
4. How to reduce the risk of exposure to the nanoparticle?

Part C   Dialogue

Dialogue 1

A: Before we discuss the reason about air pollution, I think we should clearly know what air pollution is. Air pollution is one such form that refers to the contamination of the air, irrespective of indoors or outside. A physical, biological or chemical alteration to the air in the atmosphere can be termed as pollution. It occurs when any harmful gas, dust, smoke enters into the atmosphere and makes it difficult for plants, animals and humans to survive as the air becomes dirty.

B: Well, very well. Now I clearly know what air pollution is. In the past few decades, science and technology is developing every day. Our earth has changed a lot. Among these changes, I think the air pollutions resulting from a huge number of automobiles is almost the severest.

C: Yes, I can't agree with you more. With the deepening of urbanization, more cars are needed, which will make the air pollution worse. So, what can we do for this reason? What actions should be taken?

D: First, we should apply the most cutting edge technologies in order to adopt new forms of energy as substitutes for fossil fuels. Second, try hard to develop possible transportation means which are environmental friendly. Like bicycle, battery car and even walk, so that the citizens can reduce the dependence on cars.

A: Car is a very important reason for air pollution. What about reduction of trees? I hold the view that trees play a key role in our air.

B: Yes, I agree with you. Forest, as intimate partner of mankind, is an important part of the global ecological system. If there is no forest land, the oxygen in the atmosphere will reduce, carbon dioxide will increase. The forest is a precious natural resource, the human's pillar foundation, and is known as the "lungs of the earth". The areas of forests are getting smaller and smaller. Some scientists say that there will be no vast forests in 20 or 30 years. It is really a terrible thing. We must do something to stop this bad thing. What can we do?

C: First, restore the farmland to the forest as fast as we can. We should let more people know the importance of protecting trees. Second, if we have to cut down the trees we should make a plan like trees are selectively cut on a 25 year rotation. What's more, we plant trees from now on.

D: I think industrial pollution is a big reason for bad air pollution. It was the industrial revolution that gave birth to environmental pollution as we know it today. The emergence of great factories and consumption of immense quantities of coal and other fossil fuels gave rise to unprecedented air pollution and the large volume of industrial chemical discharges added to the growing load of untreated human waste.

C: About industrial pollution, we must let government strengthen supervision.

Source: https://wenku.baidu.com/view/9e573f40c950ad02de80d4d8d15abe23492f0351.html

Dialogue 2

Jack: Good morning! Tom! How are you?
Tom: Fine, thank you! Ugh! This air smells polluted.
Jack: Yes, I have the same feeling too. I think it is because of the factories nearby.

They are polluting the surrounding air.

Tom: Nowadays, more and more people are developing in this world and starting to overuse the land and minerals of earth.

Jack: Yes, and because of this, the climate and temperature is also changing, causing global warning.

Tom: I believe that we all have a responsibility to take care of our environment.

Jack: But how?

Tom: Three things: reduce, reuse, and recycle.

Jack: Can you explain more thoroughly?

Tom: Well, for example, to reduce the amount of paper used everyday.

Jack: ...and how would you do that?

Tom: Write and print on both sides, don't write or print un-needed scripts.

Jack: So that it would be to reuse paper and other stuff?

Tom: Yes, many things can be reused over and over again, like a cardboard box can be reused and continually putted in much different stuff.

Jack: Yes, I am listening.

Tom: And finally, recycle. Recycle is somehow similar to reuse, because they both had to do with using materials over and over again, the only difference is that you send recyclable materials to the recycling center so it can be remade into new materials. Did you know that your newspapers that you read everyday are made from recycled paper?

Jack: Cool!

Source: https://wenku.baidu.com/view/0f562135a5e9856a5612607e.html

Part D　Supplementary Vocabulary

Air Pollution

acid rain　酸雨
airborne particulate matter　大气颗粒物
air pollution prevention and control　大气污染防治
ambient air quality　环境空气质量
air quality index（AQI）　空气质量指数
ambient air quality standard　环境空气质量标准
atmospheric circulation　大气环流
atmospheric cleanup　气体净化
atmospheric components　大气组分
atmospheric precipitation　大气降水
atmospheric stability　大气稳定度
anti-dust gauze mask　防尘口罩
bag filter　布袋除尘
carbon dioxide　二氧化碳

carbon tetrachloride 四氯化碳
centrifugal force 惯性离心力
chimney effect 烟囱效应
climatic zones 气候带
complex air pollution 大气复合污染物
denitrification [diːnaɪtrɪfɪ'keɪʃn] n. 脱硝作用
desulphurization [diːsʌlfəraɪ'zeɪʃn] n. 脱硫作用
detrimental [ˌdetrɪ'mentl] adj. 不利的；有害的
dry adiabatic lapse rate 干绝热递减率
dust fall 降尘
dry flue gas 干烟气
effective chimney height 有效烟囱高度
electrostatic precipitator 静电除尘器
emission concentration 排放浓度
fine particulate matter（PM2.5）空气动力学直径小于等于2.5微米的颗粒物
fixed source pollution 固定源污染
floating dust 飘尘
formaldehyde 甲醛
greenhouse gas 温室气体
haze [hez] n. 阴霾；薄雾
heat island effect 热岛效应
hydrogen fluoride 氟化氢
hydrogen sulfide 硫化氢
inhalable particulate matter（PM10）动力学直径小于等于10微米的颗粒物
isokinetic sampling 等速采样
mesosphere ['mezəsfɪə(r)] n. 中间层
methane ['miːθeɪn] n. 甲烷；沼气
mobile source pollution 移动源污染
moist plume 湿烟气
nitric oxide 一氧化氮
nitrogen dioxide 二氧化氮
oxynitride 氮氧化合物
ozone depletion 臭氧消耗
polycyclic aromatic hydrocarbon 多环芳香烃
primary pollutant 一次污染物
radioactive substance 放射性物质
sea level rise 海平面上升
secondary pollutant 二次污染物
solar radiation 太阳辐射
substitute or synthetic natural gas（SNG）替代或合成天然气
sulfur dioxide 二氧化硫
sulphur oxides 硫氧化合物
temperature inversion 逆温
thermosphere ['θɜːməsfɪə(r)] n. 热层
transpiration 蒸腾作用
volatile organic compound 光化学氧化物
wet electro- static precipitator（WESP）湿式电除尘

Part E Supplementary Knowledge

Supplementary Knowledge 1

Air Pollution Prevention and Control Action Plan

Contents

1. Increase Effort of Comprehensive Control and Reduce Emission of Multi-Pollutants.

2. Optimize the Industrial Structure; Promote Industrial Restructure.

3. Accelerate the Technology Transformation; Improve the Innovation Capability.

4. Adjust the Energy Structure and Increase the Clean Energy Supply.

5. Strengthen Environmental Thresholds and Optimize Industrial Layout.

6. Better Play the Role of Market Mechanism and Improve Environmental Economic Policies.

7. Improve Law and Regulation System; Carry on Supervision and Management Based on Law.

8. Establish the Regional Coordination Mechanism and the Integrated Regional Environmental Management.

9. Establish Monitoring and Warning System; Cope with Heavy Pollution Weather.

10. Clarify the Responsibilities of the Government, Enterprise and Society; Mobilize Public to Participate in Environmental Protection.

大气污染防治行动计划

一、加大综合治理力度，减少多污染物排放
二、调整优化产业结构，推动产业转型升级
三、加快企业技术改造，提高科技创新能力
四、加快调整能源结构，增加清洁能源供应
五、严格节能环保准入，优化产业空间布局
六、发挥市场机制作用，完善环境经济政策
七、健全法律法规体系，严格依法监督管理
八、建立区域协作机制，统筹区域环境治理
九、建立监测预警应急体系，妥善应对重污染天气
十、明确政府企业和社会的责任，动员全民参与环境保护

Supplementary Knowledge 2

如何筛选和阅读文献

一、文献筛选三步法

如何能在尽量短的时间内掌握尽量多的有价值的信息，很大程度取决于筛选文献的效率。影响因子 IF 是一个不错的线索，但一家期刊的影响因子 IF 可能会被它发过的某一篇特别优秀的文章拉高，并不见得该刊每篇都一样好，所以只是参考。

初筛是通过标题、摘要和图表来实现，思考以下几个问题：
(1) 文章回答了什么科学问题；
(2) 主要结论是什么；
(3) 什么样的证据能支持该结论；
(4) 它的实验数据是否当真支持这个结论；
(5) 证据的质量如何；
(6) 该结论重要吗？为什么重要？

一看标题：能概括关键信息最好

标题应当是对文章的精练总结，最好能明确说清作者做了什么，或得到了什么，比如发现了什么新理论、新效应，开发了什么新方法等。标题隐晦的，未必是文章不好，但标题实用的文章却也说明作者概括力强，能抓住研究重点，写作风格偏直白，文章可能更易于理解。再者，时间紧迫的时候，选一篇只从标题就能迅速判断是否契合自己所关心的问题的文章不是更好吗。看完标题可以顺便看一眼作者信息，主要是所属单位。养成看作者和机构的习惯，也会让你熟悉本领域内有哪些研究组，提高将来跟踪研究进展的效率。

二看摘要：初步判断研究质量

摘要是全文的总结，包括研究背景、目的、方法、结果和结论。通常背景和目的只要一两句话就可以了，告诉我们现在该领域里有什么、缺什么、本研究要填哪个坑。读完这两句，我们可以结合自己现有的知识先思考一下，要是我来回答这个问题，我需要哪些证据，怎么设计实验才能得到这些证据。其实这就是方法，再往下读一句就是了，可以印证或补充自己的思考，符合的就留下。如果你对自己的研究领域很有把握，觉得作者的方法有问题，那可以就此别过；如果他的方法和你所想不同，却又撩起你的探索欲，那也可保留。然后重点看看结果和结论，是否是眼下所需，对自己的课题有用；或者是否解决了本领域一个重要的疑问，推动了学科研究进展。如果你是刚开始接触一个新的领域，那可能摘要也是不太看得懂的，得先读 Introduction。如果 Introduction 仍然读得不明就里，那就先放下该文，去追踪它的参考文献，尤其是综述，先从综述中了解本研究领域的整体情况。

三看图表：从方法与结果进一步辨析

看到图表，其实已经开始获取核心资讯了，只不过再判断一下是否有必要花更多时间，去进一步读文字论述。

图表大体分两种。一种是流程图，描述本研究的方法和实验过程，也是对摘要中的 Method 部分的细化。不一定每篇文章都有，但若有则更好，从此图中看出研究过程的一些细节和数据，用以评价实验设计是否严谨。另一种是我们更为熟悉的数据图，即用可视化的手段展示实验结果，比摘要中那一小段结果更加丰富细腻有血有肉，也是文献筛选阶段的重点。图表结合 Legend 一起看，一边继续琢磨，这些数据能支持结论吗？它们之间的逻辑关系是怎样的？它们是从哪几个方面来支持结论？这些数据有没有可能换个方向去解读，得到其他结论？如果你发现数据和结论之间的逻辑链有漏洞，就无须精读。

如果明明还有其他解读方向，可能是作者没想到，也可能根本就是作者故意回避，要不要追究就看你的兴趣和时间安排了。如果数据连"们"都没有，只是单一维度，那还是放下去看看更丰满的文章吧。其实这些思考，几乎就是 Discussion 了。

读完图表，一篇文章的信息也掌握得差不多了。往后精读，就是挑几篇对眼下研究最有用的、对学科进展意义比较大的，继续读 Discussion 来印证和补充自己关于数据和结论的思考，学习别人的分析过程。然后前往 Materials and Methods 学习研究的细节，总结方法套路；再读读 Results 学习结果怎么用文字描述。

二、文献阅读的方法

每日所读之书，最好分两类，一类是精读的，一类是一般浏览的，因为我们一面要养成读书的习惯，一面要养成读书眼快的习惯，心不细则毫无所得，等于白读，眼不快则时间不够用，不能博搜资料。

1. **一般浏览法**

- 读书名标题和了解作者，加深对书文总体印象的认识。
- 读内容提要和目录，了解材料的主要内容。因为目录一般都具有论题的性质，读目录可以一目了然地掌握全书的结构布局和逻辑体系；还可以比较容易地发现重点部分和次要部分。
- 读序言和结束语，掌握全书的背景材料和结论。

2. **泛读法**

所谓博览群书，广泛涉猎。掌握的资料越多，对问题的认识也就越全面越正确；有广博的知识，可以为我们专门研究奠定坚实基础。

泛读了解表层信息和大概内容。其作用为精读作准备；用作精读之后的复读；尽快找到需要的有用的信息；开阔眼界，加大信息输入量。快速阅读法，从书中迅速吸取有用信息(每小时读速高于 3.5 万字以上)。

3. 精读法

要领：是指在粗读的基础上进行的一种深入探究，得其精髓的读书方法。

(1) 阅读时要眼到、脑到和手到。

(2) 要深入思考和分析文献的每一个概念、判断和推理的过程，以及整个文献的逻辑结构，认真把握文献的全部思想。

(3) 把文献引用、阐述的观点与客观事实及其他文献联系起来进行反复对比和研究。

(4) 对文献引用的事实、阐述的思想与研究课题之间的关系做出客观的判断和全面的评价。

"精读要一字不遗，即一个字，一个名词，一个人名、地名，一件事的原委都清楚。从头到尾地读，对照地读，反复地读。"

具体方法：

包括背读、抄读、三遍法、SQ3R 法……。理解方面，深入分析和思考；思维活动方面，采用分析、比较或联想等思维方式；记忆方面，记住主要观点和基本内容。

操作顺序：划、批、摘、评四个层次。划，边阅读边划出疑点、要点或者精彩之处；批，边读边思考，边在书上或卡片上写下个人的心得体会、见解或者疑问；摘，摘出要点；评，对书中的观点进行分析评论，对比，得出自己的观点。

背读：背读是精读法中的重要传统方法。书读百遍，其义自见，理解—熟读—强记—复读的程序。应该烂熟于心。

抄读：也是传统法，与背读结合。

三遍法：泛读(快速浏览一遍)，第二遍采用精读，第三遍再略读(跳跃式选读)。

SQ3R：美国流行的五步法。Survey(浏览)，Question(提问)，Read(精读)，Recite(复述、答疑)，Review(复习)的略语。

4. 略读法

略读法即有目的、有重点、有取舍的阅读法，在于调研与自己研究有关的信息。略读不同于精读，也不同于泛读。包括：

- 跳读：跳过无关紧要的信息，只找有益的东西。
- 楔入式：任意从某处单刀直入并向前后开拓的读法。
- 错序式：不顾文章顺序，从多处入手寻找有关信息。边阅读边思考，文章成为为自己服务的材料。
- 逆读：从后往前看，注重结果，同时猜测它的推理过程。
- 问题式：先设计问题，带着问题从文章中寻找答案。

三、积累文献资料的方法

写读书笔记最大的好处是能积累科学研究的材料，实践证明写读书笔记也是积累资料最可靠、最有效的办法。

题录型：对有价值的文章，只记录名称标题、作者姓名、出版单位或出处，需要时再深入研究。简便省事。

摘要型：摘录自己有兴趣的所有东西。事实、观点、概念、定理、名言等，忠实于原文。

评注型：写下自己阅读时的心得，哪怕是一句话。这是思想的火花，否则时过境迁就会永远消失。

导引型：对有价值的某个观点、事实，记一个大概，再记下出处(包括页码)，要求事后容易找到这个资料。

四、阅读中创造的方法技巧

1. 移植法

将别人已经发现或发明了的原理、方法、技术，移植到你所思考和研究的课题上来，从而使你所研究的问题取得进展的一种方法。就是用研究 A 对象的概念、原理、方法和成果来研究对象 B。

(1) 纵向移植：根据自然界由低级向高级依次演化发展的纵向序列，在研究中将某同一学科内部的理论中方法从低级领域(低级运动形态)向高级领域(高级运动形态)移植，或从高级领域向低级领域移植的方法。

(2) 横向移植：在研究同一层次或同一基本运动形式时，在不同学科之间进行科学原理和科学方法的移植。如日本研究《三国演义》的战争战略、战术，用于企业、商业。

(3) 综合移植：将多学科的概念、原理和方法，移植于包含多种运动形式的某一研究对象。

2. 联想法

将有关知识、原理等，通过横向或纵向的接近或相联而产生新的构思的一种方法。不同于移植。一个人是可以随时随地产生联想的。经典的联想法有三条，由古希腊亚里士多德创立：相似联想、对比联想和接近联想。

(1) 相似联想：联想到同该刺激物相似的概念或事物。

(2) 接近联想：联想到同该刺激物时间、空间接近的概念或事物，难度较大。

(3) 对比联想：联想到同该刺激物完全相反的概念或事物。
(4) 自由联想：不受任何限制的联想。有很大的成功率。
(5) 控制联想：与自由联想相对而言的，集中精力，达到目的。

3. 综合法

高度的概括和抽象。将各种信息(数据、观点和图表)加以归纳、整理，这是初级综合，融合多种因素于一体而发生质的飞跃，是高级综合。要综合运用知识；综合运用方法；综合考察对象。

4. 反向法

冲破习惯性思维的束缚而一反常规地探讨问题，能出奇制胜地解决问题。
(1) 论点反向。
(2) 论证反向，如数学中的反证法。

五、文献阅读三步法

下面就科研人员普遍喜欢采用的文献阅读三步法进行阐述。

研究者阅读文献通常因为如下几个原因：为了会议或者课堂而回顾知识；保持在本专业的前沿性；或者为研究新的领域进行文献查询。一个典型的研究者每年会在阅读学术论文上花费上百小时的时间。

学习如何高效地阅读论文这一技能很重要，但也很少会被教授。因此，学生们开始只能从自己的尝试以及错误中去吸取教训。学生在这个过程中，浪费了太多的努力，并常常有挫败感。本文阐述三步阅读法以及如何使用此方法进行文献查询。

三步阅读法最关键的一点就是你应该将论文分为三步阅读，而不是一股脑从头读到尾。每一步都在之前的基础上，完成具体的阅读目标。第一步使你把握全文的主旨思想；第二步领会论文内容，而不是细节；第三步帮助你深度理解论文内容。

1. 第一步是快速浏览全文，对全文有大体把握。

你也可以决定是否需要进行下面的两步。这个步骤应花 5 至 10 分钟，并且涵盖以下内容：
(1) 仔细阅读论文标题、摘要、简介。
(2) 阅读每一章以及每一节的开头，忽视其他部分。
(3) 如果有数学推导内容的话，扫一眼以了解背后的理论基础是什么。
(4) 阅读结论。
(5) 扫一眼参考文献，注意一下哪些你已经读过了。

在第一步的最后，应该能够回答出下列五个"C"问题：

(1) 分类(Category)：论文属于哪一类别？属于方法论，还是对现有系统的分析，抑或对研究模型的描述？

(2) 背景(Context)：论文有哪些其他的相关论文？分析问题所用的理论基础是哪一种？

(3) 正确性(Correctness)：论文所做的假设是否有理有据？

(4) 贡献(Contribution)：论文的主要贡献是什么？

(5) 明确性(Clarity)：论文写得好吗？

使用此方法，可以选择不继续深入阅读。这可能是因为论文不吸引你，或者你没办法读懂论文，或者作者做的假设无根无据。采用第一步可以使你阅读那些目前在专业以外，但可能相关的论文。

附带说一句，当你正在写论文时，你要期望大部分审阅者或阅读者对其做第一步阅读。注意选择具有连贯一致性的章节标题，并撰写简明有力的摘要。如果审阅者不能够在第一步时理解你的论文要旨，论文可能就会被拒绝；如果阅读者不能在读了 5 分钟后领会论文大意，这篇论文可能就永远不会被阅读。正因如此，一个使用精选的图画对论文做出总结的"图画式摘要"会是一种很出色的方法，并且能够在科学杂志中更多地被注意到。

2. 在第二步里，阅读论文时要更仔细，但忽视例如推导等细节。这能帮助你在阅读时草草记下要点，或者在页边进行注释。

(1) 仔细查看论文中的数据、图表以及其他插图。特别注意图表，轴线标示正确吗？结果统计上显著吗？

(2) 记得为了以后的深入阅读，标注相关的未读参考文献(这是一个了解更多论文背景的好办法)。第二步对于一个有经验的阅读者来说，应当花费一个小时左右。在这一步之后，你可以抓住论文的内容。你能够给其他人依靠支持的证据总结论文的主旨。这一细节水平适合于你感兴趣但是并不在你研究范畴内的论文。

3. 要完全读懂一篇论文，特别如果你作为一个审阅者，需要采用第三步来完成。关键在于，第三步是尝试去实际重现整篇论文：就是说，与作者做相同的假设，重新进行创造性工作。根据重现与实际论文的对比，你就能轻易分辨出论文的创新之处，以及隐藏的缺陷及假设。

Reference: http://www.sohu.com/a/128856344_170798
http://cbc.arizona.edu/classes/bioc568/papers.htm
https://wenku.baidu.com/view/80dcccff79563c1ec4da7159.html
https://wenku.baidu.com/view/3cde246b998fcc22bdd10d78.html

Unit 5   Water Pollution

Part A   Intensive Reading

Water Pollution

Pollution refers to the introduction of a substance to the environment at levels leading to lost beneficial use of a resource or degradation of the health of humans, wild life or ecosystems. Pollution may be caused by point sources at stationary locations such as an effluent pipe or nonpoint sources (also called diffuse sources) such as land runoff and the atmosphere. The mass flux of a pollutant is expressed as a load in mass per unit time.

Main Pollutants

Organic waste: organic matters in water are composed of organic compounds. Generally, most carbon-containing compounds are organic (excluding steel and simple oxides of carbon). Organic compounds include natural compounds (e.g., living organism) which are capable of decay and synthetic compounds (e.g., plastics) which may not be biodegradable. Certain organic matters in water are a source of food for aquatic microorganisms that use dissolved oxygen to convert the organic matters into energy for growth and carbon dioxide & water. The microorganism population increases in proportion to the amount of organic matters. If microbial metabolism consumes dissolved oxygen faster than atmospheric oxygen can dissolve into the water, oxygen deficit will occur and aquatic life may die when oxygen is depleted by microbial metabolism. The oxygen concentration in the river drops rapidly following organic waste pollution. Later, as the organic matter is decomposed, the oxygen resources of the river are replenished. Not all organic matters can be decomposed by microorganisms and in contrast, nearly all organic matters can be fully oxidised to carbon dioxide with a strong oxidising agent under acidic conditions.

Many organic matters are synthetic and not biodegradable (termed ***refractory organic matters***). They include pesticides, detergents and petroleum hydrocarbons. Some pesticides and other refractory organics are the principal agents responsible for the decline of certain ecosystem species and are the threats to human health.

Suspended solids: solids in water may come from mining activities, river dredging, construction (e.g., road and bridge) and soil erosion. The increased turbidity reduces light penetration and depresses photosynthesis of ecosystems. Species diversity and abundance are affected.

Nutrients: nutrients (nitrogen, phosphorus, sulphur...) in water encourages plant growth (usually in the form of algae). These nutrients come from domestic, industrial and agricultural waste. Effluent from wastewater treatment plants may still contain large concentrations of nitrogen and phosphorus albeit sufficient treatment processes have been carried out. Diffuse pollution from agricultural waste is a major source of nutrients, especially nitrogen. Nutrient enriched water is more productive with excessive biological activities and water quality is usually significantly degraded (e.g., eutrophication).

Thermal pollution: industrial cooling processes (e.g., fossil and nuclear power plants) discharge heated water into water bodies with effects on reduced solubility of O_2 and altered speed of chemical reactions. Thermal pollution changes the natural temperature regime of a water system and the affected ecosystem could be stressed. Some aquatic species may die or are forced to relocate.

Toxic metals: they are metals that form poisonous soluble compounds. Toxicity is linked with solubility and is directly proportional to its ability to cause environmental damage. Metals, particularly heavy metals are toxic (e.g., mercury, lead, cadmium, arsenic). Lighter metals may also have toxicity, such as beryllium and on the other hand, not all heavy metals are toxic (e.g., iron, which is essential for life). Toxic metal pollutions have many sources, but most commonly arise from purification of metals, mining, irrigation with wastewater, vehicle exhaust emissions.

Pathogens and carcinogens: those are disease-causing microorganisms or cancer-causing substances including bacteria, viruses, protozoa, dioxins, etc. Domestic wastewater is responsible for many of the pathogens, and industrial and agricultural wastes are the usual sources for carcinogens.

Radioactive contamination: in the natural world, there is always radiation being emitted from radionuclides. Radioactive contamination refers to unintended or undesirable presence of radiative substances in water, usually from the leakage of stored radioactive materials or release from nuclear power plant accidents, etc.

Water Pollution Indicators

The indicators on water pollution are in three categories: physical (temperature, turbidity, total suspended/dissolved solids), chemical (pH, dissolved oxygen, nitrate, COD, BOD, pesticides, metals) and biological (EPT index, Trent biotic index) indicators. Only a few indicators are explained here.

Oxygen saturation: the maximum amount of oxygen dissolved in water at a given temperature is called the saturation concentration. The oxygen demanding wastes have greater impacts on water quality in the summer as well as during dry seasons since lower river flow has less capacity to dilute the waste. Under rare occasions, actual oxygen concentration may exceed the saturation value when algae and macrophytes are photosynthesizing and producing dissolved oxygen. Such oversaturation is not stable and is eliminated when sufficient turbulence occurs. Dissolved oxygen (DO) can be measured by oxygen sensors (such as zirconia sensor, Clark-type electrode and Optodes).

Biochemical oxygen demand (BOD) is the amount of oxygen required by microorganisms to break down organic matters. It is used as an indicator of organic pollution of water. BOD is expressed in mg O_2/L and can be measured in different ways. The most common method is carried out by mixing water sample with a small amount of microorganism seeds (e.g., analogous to adding yeast to flour dough).

Chemical oxygen demand (COD) measures organic matters which are both biodegradable and non-biodegradable, hence it covers a more broad spectrum of organic matters. As a result, the COD in a water sample is always higher than the BOD. The ratio of COD/BOD provides a good indication to the proportion of organic matters that are biodegradable (COD/BOD usually varies from 1.25 to 2.50). In rivers, BOD gives a more useful estimate of the possible oxygen demand than COD and is more widely used as a measure of organic strength in the polluted water.

Biological indicators are useful to get a general reading of water quality without the high cost of lab scale analysis. Various indices are used to represent the states of macro-invertebrates. They are in two types: 1) pollution indices (existence or absence of indicator species); 2) diversity indices (to indicate the overall effect of all pollutants). However, biological indices are not always reliable. Factors other than pollution can affect the biological indices. Absence of some species may reflect its normal life cycle and quantitative sampling of biodiversity is difficult.

Wastewater Treatment

Wastewater treatment plants receive inflows from domestic, commercial, agricultural and industrial users, storm water runoff and infiltration. Industrial wastes may contain heavy metals, radioactive materials and refractory organics, which may be treated on site following specific guidelines. Domestic wastewater is commonly treated in municipal wastewater treatment plants. A main purpose for wastewater treatment is to control pollution of a receiving water body (river, lake, groundwater, sea). The most common wastewater constituents include organic matters (measured by BOD and COD), suspended solids, pathogens (disease-causing microorganisms), nutrients (nitrogen and phosphorus), toxic chemicals (heavy metals, pesticides), and others (emerging chemicals such as perfumes, pharmaceuticals).

To design an effective treatment system, it is important to 1) identify the characteristics of raw wastewater; 2) set treatment objectives; 3) integrate unit treatment operations; 4) assess the system in view of green engineering, life cycle thinking and sustainability. From Google Maps, it is quite convenient to locate a wastewater treatment plant around cities.

The treatment units include: pre-treatment, primary treatment, secondary treatment, tertiary treatment, disinfection and sludge treatment.

Pre-treatment: to remove coarse solids using screening, to remove grits using grit chambers, to remove fats and greases using flotation and to maintain a stable flow using an equalisation basin.

Primary treatment: to remove solids through gravity settling using settling (sedimentation) tanks (the removal rates are 60% suspended solids, 30% BOD, 20% phosphorus). Coagulants can be added at this stage to improve its effectiveness.

Secondary treatment: to remove dissolved organic matters using microorganisms (biological treatment). The most common biological treatment system is the activated sludge process. Effluent from primary treatment is sent to an aeration tank to be mixed with a diverse mass of microorganisms of bacteria, fungi, rotifers and protozoa. Oxygen is added to maintain the dissolved oxygen by forcing air into the system. After the process is finished, effluent is sent to the next stage and the settled sludge with microorganisms is removed from the system (with a small portion pumped to the aeration tank as microorganism seeds). At the end of this treatment, biologically degradable organic matters are decomposed and microorganisms are nutrient starved (or activated) as activated sludge.

Tertiary treatment: mainly to remove nutrients such as nitrogen (N) and phosphorus (P). It has been found that 90% of total nitrogen and 75% of the total phosphorus from household effluent are found in urine and a dual sewer system to hand separate urine and faeces would help to reduce the cost in removing them in the combined wastewater. In a wastewater treatment plant, nitrogen is removed by nitrification (ammonia → nitrite → nitrate) and denitrification (nitrate → nitrogen gas) reactions. Phosphorus can be removed either biologically or chemically. In the enhanced biological phosphorus removal, specific bacteria (polyphosphate) are used to accumulate large quantities of phosphorus within their cells and these bacteria are separated from the treated water as a fertiliser. Alternatively, phosphorus can also be removed by chemical precipitation (such as alum). Chemical phosphorus removal is easier to operate and often more reliable than biological phosphorus removal.

Disinfection: to remove pathogenic microorganisms using ozone, chlorine, ultraviolet light, or sodium hypochlorite. Unlike drinking water, any disinfectant residual is not desirable in the effluent of a wastewater treatment plant. In the UK, ultraviolet light is widely used because of the concerns about the impacts of chlorine in chlorinating residual organics in the wastewater and in chlorinating organics in the receiving water.

Sludge treatment: to treat and store the sludge from the treatment processes. To reduce sludge unpleasing odour and pathogens, sludge stabilisation is carried out by either aerobic digestion (more energy demanding due to mechanical aeration) or anaerobic digestion (more popular due to its low energy cost and methane production). Dewatering is adopted to reduce the water in the sludge before disposal. A drying bed (with an underneath drainage system) is the simplest and cost-effective dewatering method. Mechanical devices can also be used to produce a sludge cake using belt squeezing or centrifugal force.

Disposal: dewatered sludge may be buried in a landfill (carbon sequestration), incinerated (with a potential problem in air pollution), applied to agricultural land (as fertiliser). Tight control of hazardous waste from industries and households is needed to prevent harmful contamination of the sludge. Pathogens in the sludge should be controlled by heating or chemical treatment if it is applied to areas open to the public (e.g., city parks).

Natural treatment system: wetlands are an alternative wastewater treatment method. A wetland uses soil-water-air-vegetation environment to treat wastewater. Wastewater should be treated by a primary system to remove the influent solids (by settling) before it is treated in a wetland. "Living Machine" is a trademark and brand name for an ecological wastewater treatment system designed to mimic the cleansing functions of wetlands. Aquatic and wetland plants, bacteria, algae, protozoa, plankton, snails and other organisms are used in the system to provide specific cleansing or trophic functions.

Septic tank is a small-scale sewage treatment system in areas with no connection to main sewage pipes, usually in suburbs and small towns as well as rural areas. The term "septic" refers to the anaerobic bacterial environment that develops in the tank to decompose or mineralize the waste discharged into the tank. A septic tank generally consists of a tank connected to an inlet wastewater pipe at one end and a septic drain field at the other. Septic tank requires no power. A properly designed and normally operated septic system is odour-free. Waste left by the anaerobic digestion has to be removed from the septic tank by pumping, which is taken every a few years (some are up to 20 years). In areas with high population density, groundwater pollution from septic tanks may be a problem.

Source: Han D. Concise Environmental Engineering[J]. Epoca, 2012.

Words and Phrases

effluent ['efluənt] *n.* 污水，流出物
synthetic [sɪn'θetɪk] *adj.* 合成的；人造的
biodegradable [ˌbaɪəʊdɪ'greɪdəbl] *adj.* 可生物降解的
metabolism [mə'tæbəlɪzəm] *n.* 新陈代谢
detergent [dɪ'tɜːdʒənt] *n.* 洗涤剂；去垢剂；**洗衣粉**
turbidity [tɜː'bɪdətɪ] *n.* 浊度
soluble ['sɒljəbl] *adj.* 可溶的；可解的
pathogen ['pæθədʒn] *n.* 病原体
carcinogen [kɑː'sɪnədʒən] *n.* 致癌物
protozoa [ˌprəʊtə'zəʊə] *n.* 原生动物
saturation [ˌsætʃə'reɪʃn] *n.* 饱和度；饱和
macrophyte ['mækrəfaɪt] *n.* 大型植物
rotifer ['rəʊtɪfə] *n.* 轮虫

disinfectant [ˌdɪsɪn'fektənt] *n.* 消毒剂；杀菌剂
residual [rɪ'zɪdjuəl] *adj.* 残留的；剩余的
dewatering [diː'wɔːtərɪŋ] *n.* 去水（作用）；脱水（作用）
mimic ['mɪmɪk] *n.* 模仿，模拟
plankton ['plæŋktən] *n.* 浮游生物
trophic ['trəʊfɪk] *adj.* 营养的；有分泌作用的
point source 点污染
biochemical oxygen demand (**BOD**) 生化需氧量
chemical oxygen demand (**COD**) 化学需氧量
activated sludge ['slʌdʒ] 活性污泥
aeration tank [eə'reɪʃn tæŋk] 曝气池

Questions

1. How many categories of indicators on water pollution are referred in the text?
2. How to design an effective treatment system?
3. What do the treatment units include?

4. What are the main pollutants in the water?

Sentence-making

1. metabolism, disturbance, water pollution
2. point sources, non-point sources
3. chemical oxygen demand, biochemical oxygen demand, biodegradable
4. activated sludge, microbial, sediment
5. algae, protozoa, plankton,

Part B Extensive Reading

Micropollutant Fate in Wastewater Treatment: Redefining "Removal"

Percent removal is used to characterize the reduction of conventional pollutants in wastewater treatment plants (WWTPs). However, this concept of removal does not facilitate an informed discussion of micropollutants. It is increasingly apparent that for many micropollutants, transformation products (TPs) (conjugated or metabolized forms of the parent compound) exist and may even exceed the parent form in both concentration and toxicity in WWTP effluents. For example, oxidation of the antiviral drug acyclovir and its biodegradation product carboxy-acyclovir, produces a TP more acutely toxic than its parent form. TPs complicate our understanding of how treatment conditions affect micropollutant fate and are a contributing factor for large reported differences in removal of micropollutants across WWTPs. Our limited understanding of micropollutant degradation and transformation pathways prevents us from aligning the design and operation of WWTPs to reduce chemical exposure and associated risk to receiving environments. In this viewpoint, we aim to raise awareness of the misleading conclusions that may arise by describing micropollutant fate in wastewater treatment in terms of compound "removal". To more accurately interpret environmental data, focus should be shifted from removal as currently used to a comprehensive understanding of micropollutant fate that incorporates the mechanisms and kinetics of TP formation and disappearance.

A significant body of research supporting the formation and further conversion of TPs is emerging in the scientific literature. Pharmaceuticals may be biotically and abiotically transformed in the human body, sewer, WWTP, and environment. For instance, steroids are conjugated with glucuronide and sulfate moieties in the human body, making them more soluble and allowing their excretion. In WWTPs, reported observations of "negative removal" suggest that conjugated forms of micropollutants enter WWTPs and retransform, or deconjugate, back to their parent form. There is also evidence of biologically mediated conjugation of chemicals by ammonia oxidizing and heterotrophic bacteria in WWTPs, resulting in compounds that have the potential to retransform back to their parent form in the environment. Similarly, reversible abiotic transformation of sulfamethoxazole into deaminated and nitro species was observed under denitrifying conditions. If chemicals are being transformed in WWTPs, conventional reports of removal mask the true concentration of micropollutants in effluents and the overall chemical exposure in aquatic environments.

Future research must build upon emerging knowledge that affirms the existence and prevalence of TPs and delve further into understanding when, how, and why micropollutants are transformed. For example, we must better understand how redox environments, or combinations of redox environments, lead to transformations. Knowledge of common transformation pathways in redox environments may help improve the design of WWTPs in order to minimize risks associated with micropollutants and their TPs in effluents. Kinetic studies that relate transformation rates to substrate and/or electron acceptor concentrations will lead to the development of more accurate models for predicting chemical transformations. Additionally, identification of key enzymes and measurements of their relative activities can shed light on the transformation mechanisms and principal microbial metabolisms involved in transformations of micropollutants.

As it is unrealistic and unnecessary to identify every possible TP for a given micropollutant, methods must be developed to identify and prioritize prevalent TPs likely to pose risk to environmental health. Micropollutants and TPs present in effluents must be linked to potential points of exposure, which in turn must be related to risk in receiving environments. Due to the expense and difficulties in monitoring trace constituents, modeling is essential for identifying TPs and locations that are likely to experience environmental risk. As TPs that pose the greatest risk become better understood, identification of common transformation pathways for chemically related compounds will be extremely valuable in advancing structure based biodegradation prediction tools. Chemicals may be categorized based on general structures and their associated risk; for example, persistent glucuronide and sulfates, which may deconjugate back to active ingredients. Ultimately these tools can be integrated into screening and exposure assessments so that "high risk" TPs formed during treatment are identified in addition to their parent forms.

WWTP conditions that enable advantageous transformations can be harnessed to reduce chemical exposure and associated risk of micropollutants in receiving waters. As mineralization (conversion to CO_2) may be impossible for many chemicals within time frames relevant to preventing environmental exposure, a desirable alternative is to form TPs that are stable and less toxic (i.e., nonaromatic, nonchlorinated, low molecular weight compounds), or can be targeted with downstream technology. For instance, adding a downstream sorbent will not necessarily remove effluent micropollutants if the upstream treatment enables hydrophilic conjugations. However, systems that convert micropollutants to highly polar compounds can be paired with downstream adsorbents. Recognizing that process decisions may be dictated by treatment of conventional pollutants (i.e., nutrients), understanding redox environments and their role in micropollutant and TP management may help mitigate risk with limited technological or infrastructural modifications.

In order to protect the environment from the thousands of potentially toxic chemicals that pass through WWTPs, identifying and quantifying risks associated with micropollutants and their TPs are urgently needed. Focusing on the most prevalent and toxic micropollutants and TPs requires an interdisciplinary approach and collaboration between analytical chemists, environmental engineers, toxicologists, and risk assessors. Identifying and quantifying TPs in wastewater influents and effluents are first steps in moving away from potentially misleading percent "removal" values. The next essential step is to understand the conditions that influence transformation and harness technologies that enable mineralization, conversion to stable and environmentally benign TPs, or TPs that can be targeted with downstream technologies. With this knowledge, regulators will be better suited to develop policies and guidelines to control micropollutant emissions and support technologies that achieve these end points. Based on our current understanding of fate and impacts of micropollutants discussed here, it would be unwise to suggest regulating concentrations of specific chemicals. Only by rethinking "removal" and working toward a better understanding of how environmental conditions lead to desired transformations will it be possible to design and operate WWTPs to reduce risks caused by micropollutant exposure.

Source: Stadler L B, Ernstoff A S, Aga D S, et al. Micropollutant fate in wastewater treatment: redefining "removal"[J]. Environmental Science & Technology, 2012, 46(19):10485-10486.

Words and Phrases

micropollutant [maɪkrɒpə'lu:tənt] *n.* 微量污染物
[例句] The removal of organic micropollutant from drinking water has become a crucial problem.

antiviral [ˌænti'vaɪrəl] *adj.* 抗病毒的；抗滤过性病原体的；*n.* 抗病毒物质
[例句] Call your doctor to see if he or she considers you a candidate for an antiviral treatment like Tamiflu.

biodegradation ['baɪəʊdɪgreɪ'dɪʃən] *n.* [生物] 生物降解；生物降解作用
[例句] The treatment of those compounds may be classified to physical method, chemical method, physical & chemical method and biodegradation method.

kinetics [kɪ'netɪks] *n.* [力] 动力学
[例句] So now we're going to connect kinetics and thermodynamics.

sewer ['su:ə(r)] *n.* 下水道；阴沟；裁缝师；*vt.* 为…铺设污水管道；用下水道排除…的污水；*vi.* 清洗污水管
[例句] The water we use in our houses goes through the sewer and a treatment facility and back into a river or the ground where it can be re-used.

steroid ['stərɔɪd] *n.* 类固醇；[有化] 甾族化合物
[例句] If doctors suspect a woman might give birth prematurely, they might treat her with the steroid betamethasone to speed the baby's lung development before it is born.

glucuronide [glu:'kjʊərənaɪd] *n.* 葡糖苷酸
[例句] Unlike their distant human relatives, however, tree shrews quickly metabolise most of the alcohol they consume into a by-product called ethyl glucuronide (EtG).

heterotrophic [ˌhetərə'trəʊfɪk] *adj.* 非自养的
[例句] Heterotrophic bacteria eat most of the carbon-based materials in raw sewage while autotrophic bacteria consume ammonia and nitrogen compounds.

sulfamethoxazole [sʌl'fæmtɒksəzəʊl] *n.* 磺胺甲恶唑；新诺明
[例句] The predict set was absorbance data of simulated sample and compound sulfamethoxazole tablets.

substrate ['sʌbstreɪt] *n.* 底物；基底；基层；底层(等于 substratum)
[例句] Researchers are building the surfaces out of metallic materials on a substrate.

mineralization [ˌmɪnərəlaɪ'zeɪʃən] *n.* 矿化；[地质] 矿化作用；使含矿物
[例句] Carbon dioxide is bubbled into water, and then combined with mineral ions to make solid carbonate materials. Enzymes in the yeast help the mineralization process.

wastewater treatment plants (WWTPs) 污水处理厂

transformation products（TPs）转化产物，转变产物，相变产物

Questions

1. What's the difference between the meaning of removal before and after redefining?

2. Why do the conventional reports of removal mask the true concentration of micropollutants in effluents, if chemicals are being transformed in WWTPs?

3. What should we do to protect the environment from the thousands of potentially toxic chemicals that pass through WWTPs?

Part C   Dialogue

Dialogue 1

Host: The Fez River winds through the city's Medina—Fez's historic medieval center and a UNESCO World Heritage Site. Heavily contaminated, almost an open sewer, it was covered over with concrete to contain the smell; it was all but forgotten in recent decades. For much of the past 20 years, architect and engineer Aziza Chaouni has been battling to restore it. Working with the city's water department since 2007, she's now restoring and reconnecting the riverbanks with the rest of the city, while creating open, green public spaces, allowing the Medina to breathe again. At TED2014, we asked her to tell the story of this extraordinary task.

How did you begin the task of uncovering the Fez River?

Guest: The whole story actually started as my thesis at Harvard. My thesis advisor told me to do something "that you feel passionate about and that could make a difference." For years, I'd seen the river in my hometown being desecrated, polluted and filled up with trash and rats. It had become an open sewer and a massive trash yard at the core of the city.

The Fez Medina has about 250,000 inhabitants, and all their untreated sewage went straight into the narrow river that ran through it. The river was also heavily contaminated by nearby crafts workshops and tanneries — with chemicals such as chromium III, which is lethal. People working in the tanneries were getting skin cancer, and some of them were dying. It was terrible. Obviously the river started to stink, so people started building walls to block the view. Then, because it became a health hazard, they covered it with concrete starting in 2002. And because it was covered, people began using that open space as trash yard.

Actually, the first covering began in 1952, when Morocco was still a French protectorate, but it was for political reasons — so that French colonial power could easily enter the Medina and control the population. Then, as the population grew and Morocco became independent, covering happened because of the stench.

Host: In your Fellows talk at TED2014, you showed how the water fed into both public fountains and those in private courtyards. Did people actually use that water? Were they getting sick?

Guest: Of course they were, especially from the toxic chemicals dumped in the river by craftsmen. It became dangerous to drink from a running fountain. Besides, a series of droughts and excessive extraction from the water table left little water available for the Medina water network. By the 1980s, most of the fountains had become defunct, yet they had been central to its urban fabric. Imagine if Rome had no more running fountains! Can you imagine La Seine or the Thames being suddenly covered? The Fez River is smaller in scale, but the effect is similar: a central part of the city was amputated. When I witnessed all this, I was in college at Columbia University in New York at the time. I would have been 19. I was outraged; I wrote an article in the newspaper and I received hate mail, because of course it made the city look bad. At the time, I was an aspiring engineer. But due to my age and lack of experience, I was not taken seriously.

Source: http://blog.ted.com/from-an-open-sewer-to-a-jewel-of-the-city-aziza-chaouni-on-uncovering-and-restoring-the-fez-river/

Dialogue 2

A: The water in the lake smells so terrible!

B: Do you mean the pound in front of the main building?

C: Yeah, we have been there just now. The water is seriously polluted.

D: Not only the water in our school, but also in the Yangtze River and other places, the water pollution is becoming more and more serious.

A: Yes. Water pollution has now become a major global problem and is said to be the second most important environmental issue next to air pollution. Any change in the physical, chemical or biological properties of water will have a harmful effect on living things. Water pollution affects all the major water bodies of the world.

B: Besides the water pollution and air pollution, there are also many other environmental problems, relating to growth in the human population and the shortage of resources, such as white pollution, carbon dioxide emission, water and soil erosion, ozone-layer depletion and so on.

C: The environmental problems have been caused by many complex reasons. Take the water pollution as an example. Our increasing population will tremendously increase urban wastes, primarily sewage. On the other hand, increasing demands for water will decrease the amount of water available for dealing with wastes. Due to water pollution, the entire ecosystem gets disturbed.

D: And the destruction of ecosystems exacerbates the shortage of water resources further because of the decreasing of their regulating capacity. Each part of the nature is closely related. When one part is abnormal, the other will also be influenced. As we all know, the world consists of more than 70 percent water, and is covered by the atmosphere. Pollution can cause serious problems.

A: Environmental problems have been recognized as an important source of threats to human survival. Nowadays, environmental issues have caught the attention of the human race.

B: The government has also taken measures to protect the environment by establishing laws. Strengthening the environmental protection is one of the basic national policies in China.

C: In my opinion, the environmental protection is affected by several interwoven factors, including the environmental legislation, ethics and education.

D: So you mean that all of these factors have play one's part in influencing national level environmental decisions and personal level environmental values and behaviors.

C: Of course. And the government couldn't overlook any one of them.

A: I agree with your idea. I think that the awakening of people's awareness of environmental protection is the most important. It's the fundamental solution to the environmental problems.

B: And this requires the government to enhance the propaganda of environmental protections. The quality of people in the society may have a great influence.

C: As far as I am concerned, the development of science and technology plays an important role in it. For example, the development and utilization of solar energy has effectively relieved the resource shortage.

D: Have you heard of the biological control in agriculture?

A: Biological control? What does it mean?

D: It is a method of controlling pests, such as insects, mites, weeds and plant diseases. What is different from the traditional method is that it relies on predation, parasitism, herbivory, or other natural mechanisms.

C: Biological control can not only reduce the use of pesticides, which is harmful to both the environment and our health, but also have a higher efficiency.

B: The environmental protection is not only the duty of one nation, but also all the human beings. The international cooperation is also necessary. Now we have made much progress in this area.

A: Yes. For example, the signing of Kyoto Protocol has made the program, that the world takes action to reduce carbon emissions which are causing the huge and destructive climate change.

C: But there are also many difficulties in the implementation of policy and program. What is the most important is to make a balance between the economy and environment. Or they will be hard to received people's approval when they decrease people's living level.

B: Anyway, although complete environmental protection seems impossible at this current global position, many countries have established organizations and agencies devoted to environmental protection. There are also many international environmental protection organizations.

D: So when we want to make contribution to the environmental protection, we can take part in such organizations and be a volunteer.

A: It's the duty for everyone to protect our home. Not only can we participate in some organizations, but also take actions in our daily life. For example, we can reduce the use of plastic goods, classify retrieving garbage and so on.

B: We should also pay attention to saving the energy. Turn the light off when you leave, turn off the air conditioner when nobody in the room, and the reuse of the water…These all are the good habits in everyday life.

C: Umm…I have an idea about the polluted pound in front of the main building.

D: What?

C: We can write a letter about the pollution of water body in our school to the headmaster to recommend treating the sewage.

D: And first we should make a survey. With the specific statistics our advice will be more convincing.

A: That's a good idea. Let's go!

Source: https://wenku.baidu.com/view/dec2a8b9d5bbfd0a7956737b.html

Part D Supplementary Vocabulary

Water Pollution

activated carbon adsorption 活性炭吸附
activated sludge process 活性污泥法
adjusting tank 调节池
advanced treatment 高级氧化处理
aerobic 好氧的
aerobic treatment 好氧处理
algae 藻类
anaerobic 厌氧的
anaerobic anoxic oxic process（AAO）厌氧缺氧好氧的过程
anaerobic treatment 厌氧处理
anoxic 缺氧的
anoxic oxic process（A/O）前置缺氧-好氧工艺
bio-chemical oxygen demand（BOD）生化需氧量
bio-chemical process 生物化学过程
biofilm process/bio-membrane process 生物膜法
biological contact 生物接触氧化
biological reactor 生物反应池
biological treatment 生物处理
biological water-quality index 生物性水质指标
bioremediation 生物修复
biotransformation 生物转化
carbonate 碳酸盐
catchment area 排水区，下游区
chelate-induced 螯合诱导
chemical oxygen demand（COD）化学需氧量
chemical treatment 化学处理
chemical water-quality index 化学性水质指标
chloride 氯化物
coagulate flocculant 混凝剂
coagulate flocculating agent 混凝沉淀
colloidal material 胶体物质
confined water 承压水
cyanide 氰化物
cyclic activates sludge technology（CAST）循环式活性污泥工艺
desalination 脱盐
dilute 稀释
dissolved oxygen（DO）溶解氧
dissolved solids（DS）溶解性固体
domestic sewage 生活污水
domestic waste water 生活污水
dysentery 痢疾
flocculant 絮凝剂
flooding 泛滥的
flotation 气浮
freshwater 淡水；内河；淡水湖
freshwater degradation 淡水恶化
groundwater 地下水
hardness 硬度
industrial wastewater 工业废水
microbe 微生物
mixed liquor suspended solids（MLSS）混合液悬浮固体浓度
municipal wastewater 城市生活污水

mixed liquor volatile suspended solids（**MLVSS**）混合液挥发性悬浮固体浓度
outfall 河口，排水口
oxidation ditch 氧化沟
oxidation pond 好氧塘
pathogen 病菌
total phosphorus（**TP**）总磷
physical-chemical treatment 物理化学处理
physical equipment 加药设备
physical indicate of water quality 物理性水质指标
physical treatment 物理处理
point source 点源
ponds tailing 池塘尾渣
primary sedimentation tank 初沉池
primary treatment 一级处理
protein 蛋白质
refractory 难降解的
remediation goal 整治目标
runoff 径流
screening 过滤
secondary sedimentation tank 二沉池
secondary treatment 二级处理
sedimentation 沉淀
sedimentation tank 沉淀池
setting velocity（**SV**）污泥沉降比
sludge bulking 污泥膨胀
sludge volume index（**SVI**）污泥容积指数
source apportionment 源解析
stabilization pond 稳定塘

subterranean water 地下水
suspended substance（**SS**）悬浮物
terrain 地形
tertiary treatment 三级处理
total dissolved solids（**TDS**）总溶解固体
total nitrogen（**TN**）总氮
total organic carbon（**TOC**）总有机碳
total suspended solids（**TSS**）悬浮固体总量
trickling filter 生物滤池
turbidity 浑浊度
typhoid 伤寒
undegradable 不可降解的
upflow anaerobic sludge blanket（**UASB**）流式厌氧污泥床
virus 病毒
volatile 易挥发的；爆炸性的
waste water 废水
water body 水体
water pollutant 水污染物
water pump 水泵
water quality purification 水质净化
water quality standard 水质标准
water quality 水质
water resource 水资源
water salination 水盐碱化
watershed 流域
watershed management 流域管理
water supply 供水
water treatment 水处理
wetland ecosystem 湿地生态系统

Part E Supplementary Knowledge

Supplementary Knowledge 1

The Action Plan for Prevention and Control of Water Pollution

Contents
1. Overall Control of Pollutant Discharge
2. Promotion of Transformation and Updating of Economic Structure
3. Focus on Water Resources Saving and Conservation
4. Strengthening of Sci-Tech Support
5. Give Full Play to the Function of Market Mechanism
6. Tightening of Environmental Law Enforcement and Supervision
7. Effective Strengthening of Water Environmental Management
8. Full Guarantee of Water Ecological Environment Safety
9. Defining and Fulfilling the Responsibilities of Each Party
10. Strengthening of Public Participation and Social Supervision

水污染防治行动计划

一、全面控制污染物排放
二、推动经济结构转型升级
三、着力节约保护水资源
四、强化科技支撑
五、充分发挥市场机制作用
六、严格环境执法监管
七、切实加强水环境管理
八、全力保障水生态环境安全
九、明确和落实各方责任
十、强化公众参与和社会监督

Supplementary Knowledge 2

学术演讲的内容和结构

内容
为了在规定的时间内做一次精彩演讲，你需要：
- 选择最重要的部分来讲
- 用清晰、有条理、有视觉吸引力的幻灯片展示演讲脉络

- 用速度稍慢的简单易懂的英语解释每张幻灯片

演讲的内容与论文的内容类似，包括引言、结果（与实验方法综述相结合）、讨论和结论。然而，作为口头形式的演讲与书面形式的论文在结构和措辞上有所不同。如果你想带给观众一场精彩的演讲，就不要照着论文逐字读出来。你需要了解读者和听众是如何以不同的方式提炼信息的。对于读者来说，他们只能跟随演讲者的节奏，并且只有一次机会听到某一条信息。相对地，听众不能控制他们即将看到和听到的信息的顺序和类型，然而读者可以在需要的时候轻易浏览到标题、副标题，或者跳过某一段信息。另外，在演讲时，演说者较容易表达自己的感情；然而书面形式则难以做到这点。

不同长度演讲的设计

无论你是在会议上花10分钟做简短的演讲，还是在研讨会上花一个小时演讲，演讲在大体结构上都需要有引言、结果和讨论这些板块。根据目标观众的不同，可能需要使用不同数量的介绍性幻灯片。对于非科研人员来说，最好提供更多关于背景知识的幻灯片，因为大多数演讲者会在最初的几分钟里因为对研究问题的不恰当介绍而失去观众。

与较长的演讲相比，简短的演讲一般更难准备和进行。在简短的演讲中，你需要非常有选择性，选出每个部分最重要的信息，并且在准备幻灯片的时候牢记这些重点。通常情况下，在简短的演讲中，你需要减少背景知识幻灯片的数量，把精力集中在研究的主要发现和对它们的阐释上。实际上，根据主题和观众的差异，你可能只有展示1~2个主要发现的时间。

当时间很紧的时候，可以考虑跳过只有标题和内容概览的幻灯片。作为补偿，在口头上告知观众演讲的总标题和内容概览。在任何情况下都要有一张结论/总结的幻灯片。

总体结构

口头报告指南1：演讲总体结构

告诉观众你将要讲什么

向观众讲述

告诉观众你刚刚讲了什么

口头报告指南2：按照以下提示组织幻灯片

可选：	标题幻灯片
第一张：	内容概览
第二张：	引言和背景
接下来几张：	展示研究对象、研究方法和结果
最后一张：	展示结论和支持结论的重要证据
可选最后一张：	感谢同事和研究的资助机构

在演讲开头，告诉观众你将要讲什么，也就是演讲的纲要：介绍主题，告知观众你将要如何展示你的研究成果，然后将总结本次演讲的内容。

在介绍完演讲的纲要之后，介绍研究主题的背景并引申到你的研究，即已有的研究尚未解决的问题，随后介绍你的研究目的/问题。在介绍研究结果时简要描述实验方法，并在整个研究背景下讨论研究发现。在最后一张幻灯片上，写上阐释和结论。演讲结束时，向观众表示感谢。

作为演讲者，注意不要用头、手或身体的其他部位挡住投影。另外，注意不要挡住任何人的视线。

视觉辅助的形式

　　口头报告指南3：在演讲之前提早准备视觉辅助资料
　　口头报告指南4：使它们看起来有吸引力
　　口头报告指南5：使它们简单易懂
　　口头报告指南6：以图像的形式来思考

在开会前提早准备视觉辅助资料以便自己有时间来检查、替换或改进它们。无论你使用头顶投影机、幻灯片或 PowerPoint，都要简单。幻灯片必须清晰、易读、易懂。避免使用不标准的色彩、字体、图表和缩写。这些视觉辅助资料应当有助于宣传你的研究工作而不是转移观众的注意力。它们应当使观众获得视觉上的愉悦感，但是又不失专业性。此外，观众应该可以辨认出每一张幻灯片所属的部分(相同的色彩、字体风格和强调手段)。

形式和色彩

在演讲中，清晰的图像形式的视觉辅助手段是不可替代的。为制作有视觉吸引力的幻灯片，应留意其他人是怎样使用易被眼睛接受的色彩、字体和图表的。避免使用明亮的颜色和红色/绿色(或蓝色/橘红色)这样的对比色，因为有些人是色盲不能辨认这些颜色。在背景和内容或图片之间要有较高的对比度，例如深色文字配以浅色背景或相反。中蓝或深蓝色背景配以白色或黄色内容是经常被使用的搭配，因为背景和内容的对比度大，它们非常易读。

要想用尽可能少的文字制作有效的幻灯片，了解如何用分项的形式简要地表达意思十分重要。充斥文字的幻灯片通常能为演讲者带来便利却不能给观众带来便利。一定要避免使用这类幻灯片。为了使幻灯片更吸引观众：①每张幻灯片最多使用 40 个词，每行最多 40 个字，全是文字的幻灯片最多不超过 14 行；②使用关键词和词组(它们比句子更有效果)；③避免以数字为编号罗列项目，应该使用项目符号，每一项尽量只用 3~5 个词，每张幻灯片不超过 7 个项目；④尽量少使用标点符号，并从幻灯片左侧而不是中间开始一段文字。

需要注意的是文字的字号要足够大，确保不使用小于 18 磅的字体。使用无衬线字体，例如 Arial，它们比衬线字体更易读，例如 Times New Roman。使用小写字母，必要的时候首字母大写比全部使用大写字母更易读。用最大对比度和黑体来达到最大的易读性。幻灯片上的每个字都应该是观众的目标语言。用色彩和(或)照片图片这样的视觉手段在适当的地方进行强调。最重要的信息用大号字体显示，斜体也可用来进行强调，但是避免使用下划线和惊叹号，因为人们并不认为它们很有效。

图表

每张幻灯片的图表上方都应该有标题，标题应与图表相隔一定的空间。图表中不同组数据要用名称而不是数字或字母来区分。但是，不要简单地照搬发表过的论文中的图表。这些图表的字体都很小，没有标题，而且多半都包含幻灯片并不需要的额外信息。

在图和表中首选图。条形图因其所传达的信息能被很快地理解而受到欢迎。在条形图中，控制条状物的数量。如果可以不使用表格，尽量不用。如果你必须使用表格，包括标题和项目最多不要超过4列，7~8行。所有文字都横向书写，如果可能的话，纵坐标轴的标签页横向书写。图片中的曲线要平滑，字体要清晰(即使用 Arial 或 Helvetica)，符号要容易辨别。不要将图注放入幻灯片，但是如果需要，为图片提供一个图例。确保图表中的各项都排列整齐，各自之间和与之相关的文字互为照应。

演讲准备

在演讲开始之前，应该检查设备，并进行安装调试。让自己熟悉演讲的环境和设备，检查照明、电源插座、黑板和放映设备。如果你的演讲要使用 PowerPoint，而且不使用自己的电脑放映，要保证你的 PowerPoint 文件能跟放映用的电脑和投影仪兼容。保证笔记的顺序正确。最后，很重要的一点是，记住要介绍你的姓名并正确发音。

声音和讲话方式

    口头报告指南 7：保证所有观众都能听见你的声音
    口头报告指南 8：讲话速度不要太快或太慢
    口头报告指南 9：避免可怕的口头语

对于英语这种语言来说，重音非常关键。虽然发音也很重要，但是使用正确的重音更重要。如果重音用得对，你不用担心有一些口音。如果你有发音困难，这样做也许会对你有帮助：在需要重读的单词音节上标注重音符号，在需要停顿的地方做标注，也可以划出需要强调的词组。

保证所有观众都能听见你的声音。讲话不要太柔和，这样会显得你不是很确定。但是，也不要用太大的声音让观众震耳欲聋。注意你的音高。声音太高可能导致观众觉得演讲者不是很专业而且孩子气。深沉、饱满的音调让你的声音更富有磁性。通过使用丹田气和下喉部位置发音可以控制音高，不要用上喉部和鼻腔发音。

讲话速度不要太快或太慢。好的演讲的速度应该比正常谈话的速度稍慢而且更清楚。紧张的演讲者常常会讲得太快。尝试使用从容不迫的节奏来讲话。在讲话时聆听自己，用自己能理解的速度讲话。

词语的选择

口头报告指南10：使用缓和语气和过渡的连接词

口语不同于书面语。演讲时，你需要缓和语气和表示过渡的词语，这些词语在最后发表的研究论文中通常被删掉了。这些过渡性词语要用在演讲里而不是幻灯片上。

将英语作为第二语言的演讲者尤其要注意书面语和口语之间的差异。当你听英语为母语的演讲者演讲时，可以考虑为自己准备一份清单，列出那些过渡性词语。但是，不要过多使用这些缓和语气的词语。

与过渡性和缓和语气的词语一样，引导性的词和词组在演讲中也很常用，但是不用在书面语中。引导性词语和词组如：I am going to present…; I would like to…; What I am going to talk about is…; To start, …; This talk is about…; My presentation deals with..; Then I am going to discuss…; At last, I would like to…; Now we move on to…; I want to spend some time on…; For the rest of the time…; On the next slide…

问题和回答

在科技会议上，口头报告之后通常会有一个简短的提问环节，任何观众都可以向展示者提问。大多数展示者对于这个环节都感到紧张，尤其是在被问到它们可能不知道答案的问题时。对于会议语言不是母语的人来说，提问环节尤为令人害怕。以下内容会帮助你减轻担忧的情绪。

在演讲之前就预想观众会提什么问题。尝试从观众的角度来看你的演讲，预想会有哪些问题。与你的同行、首席研究员或同事练习提问与问答。练习的最好方式是，在一个安静的会议室内，站着演讲，用上幻灯片(即使还未全部完成)、投影仪和激光笔；简而言之，假想是在观众面前演讲。

Reference: Angelika H. Hofmann. 科技写作与交流-期刊论文、基金申请书及会议讲演[M]. 任胜利, 莫京, 安瑞, 等, 译. 北京：科学出版社, 2012.

Unit 6 Soil Formation and Erosion

Part A Intensive Reading

Weathering

No sooner has a rock formed than it becomes vulnerable to attack by weathering. The word "weathering" is slightly misleading. We associate it with wind, water, freezing, and thawing. These are important agents of weathering, but they are not the only ones. Weathering can be chemical as well as physical and it often begins below ground, completely isolated from the weather.

Beneath the surface, natural pores and fissures in rocks are penetrated by air, containing oxygen and carbon dioxide, and by water into which a wide variety of compounds have dissolved to make an acid solution. Depending on their chemical composition, rock minerals may dissolve or be affected by oxidation, hydration, or hydrolysis. Oxidation is a reaction in which atoms bond with oxygen or lose electrons (and other atoms gain them, and are said to be "reduced"). Hydration is the bonding of water to another molecule to produce a hydrated compound; for example, the mineral gypsum ($CaSO_4 \cdot 2H_2O$) results from the hydration of anhydrite ($CaSO_4$). Hydrolysis (lysis, from the Greek *lusis*, "loosening") is a reaction in which some parts of a molecule react with hydrogen ions and other parts with hydroxyl (OH) ions, both derived from water, and this splits the molecule into two or more parts.

The result of chemical weathering can be seen in the limestone pavements found in several parts of England, Wales, and Ireland. South Devon, England, is famous for its red sandstones, well exposed in the coastal cliffs of the Torbay area. These date from the Devonian Period, some 400 million years ago, when what is now Devon was a hot, arid desert. The desert sand contained some iron, which was oxidized to its insoluble red oxide, giving the sandstone its present colour.

Iron oxidizes readily and this form of weathering has produced hematite (Fe_2O_3), one of the most important iron ore minerals, some of which occurs in banded ironstone formations, 2~3 billion years old, composed of alternating bands of hematite and chert (SiO_2). Iron and other metals can also be concentrated by hydrothermal, or metasomatic, processes. Near mid-ocean ridges, where new basalt is being erupted onto the sea bed, iron, manganese, and some other metals tend to separate from the molten rock and are then oxidized and precipitated, where particles grow to form nodules, sometimes called "manganese nodules" because this is often the most abundant metal in them. Vast fields of nodules, containing zinc, lead, copper, nickel, cobalt, silver, gold, and other metals as well as manganese and iron, have been found on the floor of all the oceans. A few years ago serious consideration was given to the possibility of dredging for them, but at present metals can be obtained more cheaply by conventional mining on land.

Hydrothermal weathering, in which hot solutions rise from beneath and react with the rocks they encounter, produces a range of commercially valuable minerals, perhaps the best known of which is kaolin, or "china clay". This material was first discovered in China in 500 BC and was used to make fine porcelain, hence the names "china clay" and "kaolin", from kao ling, meaning "high ridge", the type of landscape in which it occurred. Today it is still used in white ceramics, but most is used as a filler and whitener, especially in paper. Kaolin deposits occur in several countries, but the most extensively mined ones are in Cornwall and Devon, Britain.

Kaolin is a hydrated aluminium silicate, $Al_2O_3 \cdot 2SiO_2 \cdot 2H_2O$, obtained from the mineral kaolinite. The British deposits occur in association with the granite batholiths and bosses intruded during the Hercynian orogeny. Granites consist of quartz crystals, mica, and feldspars. Feldspars are variable in composition. All are aluminium silicates, those associated with the kaolinite deposits being plagioclase feldspars, relatively rich in sodium. As the intruded granite was cooling, it was successively exposed to steam, boron, fluorine, and vaporized tin. The feldspar reacted with these, converting it into kaolinite (the process is known as kaolinization), a substance consisting of minute white hexagonal plates which are separated from the rock industrially by washing and precipitation, leaving a residue of quartz grains (a white sand) and mica. About 15 percent of the material is recovered as kaolin, 10 percent is mica waste, and 75 percent is sand, which is also waste although it has found some use for building and landscaping. In some places the kaolinization process has been completed from above, possibly by humic or other acids from overlying organic material, but most of the kaolinite formed at depth is overlain by unaffected granite, probably because the upward movement of acidic fluids was halted by the absence of veins or joints it could attack. The resulting deposits are funnel-shaped, extending in places to depths of more than 300 m.

Bauxite, the most important ore of aluminium, is also produced by the chemical weathering of feldspars, in this case by hydration. Bauxite is a mixture of hydrous aluminium oxides and hydroxides with various metals as impurities; to be suitable for mining it should contain 25~30 percent of aluminium oxide.

Bauxite is a variety of laterite, one product of the kind of extreme weathering of soil called "laterization". The word "laterite" is from the Latin *later*, meaning "brick", and laterite is brick-hard. Laterization occurs only in some parts of the seasonal tropics, where soils are derived from granite parent material, but it is possible that removing the forest or other natural vegetation in such areas may trigger the formation of laterites. These can be broken by ploughing.

Except on steep slopes, tropical soils overlying granite can be up to 30 m deep. Naturally acidic water from the surface percolates through them, steadily eating away at the parent rock beneath, and plants draw the water up again through their roots. Water is also drawn upward by capillary attraction through tiny spaces between soil particles and evaporates from the surface. If the rainfall is fairly constant through the year, the movement of water is also constant, but if it is strongly seasonal, evaporation exceeds precipitation during the dry season and mineral compounds dissolved in the soil water are precipitated, the least soluble being precipitated first. Provided vegetation cover is adequate, with roots penetrating deep into the soil, the minerals will not accumulate in particular places and when the rains return they will be washed away. If there is little plant cover, however, they may accumulate near the surface. The most insoluble minerals are hydroxides of iron and aluminium (kaolinite) and they are what give many tropical soils their typically red or yellow colour. Soil developed over granite will contain sand, or quartz grains, and clays derived from feldspars in varying amounts. Laterite layers or nodules are hard, but not usually thick, because, being impermeable, they prevent further percolation of water downwards into the soil and thus bring the laterization process to an end. Erosion of the surface layer may then expose the laterite.

Laterization does not necessarily render a soil useless and many relatively laterized soils are cultivated, although some soils resembling lateritic soils, for example in parts of the eastern United States, are not truly laterized. Indeed, there are doubts about the extent to which laterization is occurring at present. Lateritic soils in the West Indies, Indonesia, Australia, India, and China may well be of ancient origin.

Living organisms contribute to weathering. By moving through soil they assist the penetration of air and water, and the decomposition of organic material releases acids and carbon dioxide, some of which dissolves into the soil water. Biological activity contributes greatly to the formation of soil.

Physical weathering is also important in soil formation, especially in its initial stages, but it can also degrade soils through erosion. Thermal weathering, which is the expansion and contraction due to repeated heating and cooling, causes rocks to flake, especially if water is held within small crevices. Small particles detached from the rock may then be carried by the wind and if they strike other rocks more particles may be chipped from them. Depending on their sizes, the particles may be carried well clear of the ground or may roll and bounce along the surface; the process is called "saltation". Most serious erosion is due to water, however. All water flowing across the land surface carries soil particles with it. This can lead to the formation of rills and gullies into which more particles are washed and then transported, or where water flows as sheets whole surface layers can be removed. In addition to this, all rivers erode their banks, and waves erode the shores of lakes and the sea.

These processes are entirely natural and part of the cycles by which originally igneous rocks are converted into sediments and landforms are made and age, but human activities can accelerate them. The UN estimates that in the world as a whole, some 1.093 billion hectares (ha) of land have been degraded by water erosion, 920 million ha by sheet and surface erosion and a further 173 million ha by the development of rills and gullies. Of the total area subject to serious degradation by water, 43 percent is attributed to the removal of natural vegetation and deforestation, 29 percent to over-grazing, 24 percent to poor farming practices, such as the use of machinery that is too heavy for the soil structure to support and the cultivation of steep slopes, and 4 percent to the over-exploitation of vegetation. There is, however, some evidence that modern farming techniques can reduce soil erosion substantially. A study of a site in Wisconsin found that erosion in the period 1975~1993 was only 6 percent of the rate in the 1930s. This may be due to higher yields from the best land, combined with methods of tillage designed to minimize erosion.

Weathering is the general name given to a variety of natural processes by which rock is recycled and soil and landscapes created. It creates and alters environments, but human activities can accelerate it on vulnerable land, degrading natural habitats and reducing agricultural productivity.

Source: Allaby M. Basics of Environmental Science, 2nd Edition[M].The Taylor & Francis e-Library, 2002.

Words and Phrases

weathering ['weðərɪŋ]　　*n.*　　[地质]
风化

fissure ['fɪʃə(r)]　　*n.* 裂缝；裂缝
penetrate ['penɪtreɪt]　　*vt.*　　渗透；

穿透；洞察；*vi.* 渗透；刺入；看透
hydration [haɪ'dreɪʃn]   *n.* [化学] 水合作用
gypsum ['dʒɪpsəm]   *n.* 石膏
anhydrite [æn'haɪdraɪt]   *n.* [矿物] 硬石膏；[建] 无水石膏
hydroxyl [haɪ'drɒksɪl]   *n.* 羟基,氢氧基
limestone ['laɪmstəʊn]   *n.* [岩]石灰岩
hematite ['hemətaɪt]   *n.* [矿物]赤铁矿
chert [tʃɜːt]   *n.* [岩] 燧石；黑硅石；角岩
hydrothermal [ˌhaɪdrə'θɜːməl]   *adj.*[地质] 热液的；热水的
metasomatic [metəsəʊ'mætɪk]   *adj.* 交代的
basalt ['bæsɔːlt]   *n.* [岩] 玄武岩；黑陶器
manganese ['mæŋɡəniːz]   *n.*[化学] 锰
nodule ['nɒdjuːl]   *n.* 小瘤；节结(尤指植物上的)
dredge [dredʒ]   *n.* 挖泥；清淤；*v.* 疏浚；挖掘
granite ['ɡrænɪt]   *n.* 花岗岩；花岗石
batholith ['bæθəlɪθ]   *n.* [地质] 岩基；底盘(等于 batholite)
mica ['maɪkə]   *n.* [矿物] 云母

feldspar ['feldspɑː(r)]   *n.* [矿物] 长石
plagioclase ['pleɪdʒɪkleɪs]   *n.* 斜长石
boron ['bɔːrɒn]   *n.* [化学] 硼
tin [tɪn]   *n.* 锡；罐头，罐；马口铁；
kaolinization [keɪəlɪ'naɪzeɪʃn]   *n.* [地质] 高岭石化；[地质] 高岭土化作用
hexagonal [heks'æɡənl]   *adj.* 六边的，六角形的
vein [ven]   *n.* [地质] 岩脉；纹理；翅脉
bauxite ['bɔːksaɪt]   *n.* 矾土，[矿物] 铁铝氧石；[矿物] 铝土矿
laterite ['lætəraɪt]   *n.* 红土带；砖红壤；铁矾土
laterization [ˌlætərɪ'zeɪʃən]   *n.* 红土化作用
percolation [ˌpɜːkə'leɪʃn]   *n.* 过滤；浸透
sediment ['sedɪmənt]   *n.* [地质] 沉积物；*v.* 沉积
hercynian orogeny   海戊山运动；海西期造山运动
capillary attraction   毛细作用，毛细吸力

Questions

1. How many kinds of weathering are mentioned in the text? What are they?
2. What are the main reactions of chemical weathering?
3. What is the process of laterite?
4. What is the effect of farming on soil erosion?
5. Please briefly describe the weathering effects of the organism on the rock.

Sentence-making

1. pores, rain, rock
2. wind, water, soil erosion
3. capillary attraction, evaporation, rainfall
4. soil, surface, physical attack
5. microbial, decomposition, organic material

Part B Extensive Reading

Soil Formation, Ageing, and Taxonomy

From the moment it is exposed at the surface, rock is subjected to persistent physical attack. Water fills small fissures and when it freezes it expands, exerting a pressure of up to 146 kg/cm^2, which is sufficient to split the toughest rock. In summer, the rock warms during the day and cools again at night, expanding as it warms and contracting as it cools, but it is heated unevenly. The surface is heated more strongly than rock beneath the surface; some parts of the surface are exposed directly to sunlight, others are in shade. As a consequence, some parts of the rock expand and contract more than others. This, too, causes the rock to break. Often flakes are loosened or detached from the surface, a process called "exfoliation". Detached particles then grind against one another as they are moved by gravity, wind, or water. This breaks them into still smaller pieces.

The smaller any physical object, the greater its surface area in relation to its volume: a sphere with a diameter of 4 units has a surface area of 50 units^2 and volume of 33.5 units^3, giving an area: volume ratio of 1:0.7; if the diameter is 2, the surface area is 12.5 units^2, volume 4.2 units^3 and ratio 1:0.3. As the rock particles grow smaller, therefore, the total surface area exposed to attack increases. Still vulnerable to abrasion, they are now subject to chemical attack.

This takes several forms. Some of the chemical compounds of which they are composed may be soluble in water; wetting dissolves and drainage removes them. Other compounds may react chemically with water. The process is called "hydrolysis" and can convert insoluble compounds to more soluble ones. Orthoclase feldspar ($KAlSi_3O_8$), for example, a common constituent of igneous rocks, hydrolyses to a partly soluble clay ($HAlSi_3O_8$) and very soluble potassium hydroxide (KOH) by the reaction:

Unit 6 Soil Formation and Erosion

$$KAlSi_3O_8 + H_2O \rightarrow HAlSi_3O_8 + KOH$$

Hydration is the process in which compounds combine with water, but do not react chemically with it. The addition of water to a compound's molecules makes them bigger and softer and so increases their vulnerability to breakage. Oxidation also increases the size and softness of many mineral molecules and may also alter their electrical charge in ways that make them react more readily with water or weak acids. Reduction, which occurs where oxygen is in short supply, also alters the electrical charge on molecules and may reduce their size.

Compounds may also react with carbonic acid (H_2CO_3), formed when carbon dioxide dissolves in water. This reaction, called "carbonation", forms soluble bicarbonates. Barely soluble calcium carbonate $(CaCO_3)$, for example, becomes highly soluble calcium bicarbonate $(Ca(HCO_3)_2)$.

Physical and chemical processes thus combine to alter radically the structure and chemical composition of surface rock. How long it takes for solid rock to be converted into a layer of small mineral particles depends on the character of the original rock and the extent of its exposure; in arid climates it proceeds more slowly than in moist ones, for example. Yet the process is remorseless. At widely varying speeds it dismantles mountains.

It does not proceed far before living organisms accelerate it: respiration and the decomposition of plant remains are the main source of the carbon dioxide engaged in subsurface carbonation. The chemical changes release compounds useful to organisms in soluble forms they can absorb, and their metabolic wastes and dead cells add to the stock of reactive compounds as well as providing sustenance to still more organisms. Bacteria are usually the first to arrive, forming colonies in sheltered cracks, invisible to the naked eye. Lichens often follow, composite organisms comprising a fungus and alga or cyanobacterium. The fungus obtains water and mineral nutrients from the rock, the alga or cyanobacterium supplies carbohydrates that it photosynthesizes and oxygen as a by-product of photosynthesis. Each partner supplies the other and the fungus protects them both from drying out and provides firm attachment to the rock, which it grips tightly with filaments that grow into the tiniest crevices. This remarkable partnership allows lichens to flourish where no plant could survive.

Organic material, derived from wastes and the decay of dead cells, accumulates beneath the lichen, mixing with the mineral particles and accelerating chemical reactions. This mixture is better at absorbing and holding water, and in time there is enough of it to provide anchorage and nutrients for plants. Mosses may arrive and small herbs may root themselves in the deeper cracks.

As the layer of mixed organic and mineral material thickens, some of it begins to be washed to deeper levels, a few centimetres below the surface. The material is starting to form two distinct layers: an upper layer from which soluble compounds and particles are being washed (the technical term is "leached") and a lower layer in which they are accumulating. This is the first stage in the formation of soil.

From this point, vegetation becomes part of the developing soil and contributes greatly to its formation. Plant roots penetrate the material and when they decay leave channels that assist aeration and drainage. Dead plant material contributes fresh organic matter to the surface, which decays to release compounds that drain into the soil. In detail, however, this process can vary widely over a small area, in large part because of the efficiency with which the soil drains and the depth of the water table below the surface. If the soil is derived from similar mineral particles all the way down a slope, a hydrologic sequence may occur. Where drainage is excessive, the soil will be generally dry, favouring trees that can root to considerable depth. As the distance narrows between the surface and water table, the region hospitable to plant roots becomes shallower, and the plants smaller. Most of the organisms engaged in the decomposition of plant material require oxygen for respiration, so the decreasing depth of the aerated zone is accompanied by a slowing of the rate of decomposition until, where the soil is waterlogged, partly decomposed material may form acid peat.

The stage of this process that involves purely physical and chemical mechanisms constitutes weathering; as living organisms become the predominant agents it is called "pedogenesis". Plants growing at the surface penetrate the soil with their roots and supply a topmost layer of dead organic material, called "litter". This provides sustenance for a diverse population of animals, complete with their predators and parasites, fungi, and bacteria. These break down the material, which enters the soil proper, much of it carried below by earthworms, where it feeds another population. Compounds released by decomposition dissolve in water draining through the soil and are carried to a lower level, where they accumulate. At the base of this layer, the "subsoil", rocks and mineral particles, detached from the underlying rock, are being weathered, and below this layer lies the bedrock itself.

If a vertical section, called a "profile", is cut through the soil from surface to bedrock, it may reveal this structure as layers, called "horizons", clearly differentiated by their colour and texture. But in any particular soil there may be more or fewer and in some soils horizons are not easily distinguished at all. Conventionally, the horizons are identified by letters: O for the surface layer of organic matter; A for the surface horizons; B for the accumulation layer; C for the weathering layer; and R for the bedrock (Figure 1). The horizons are classified further by the addition of numbers:

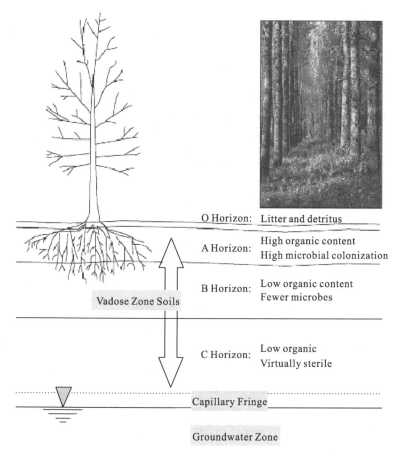

Figure 1 Soil horizons
(Source: Vaccari D A, Strom P F, Alleman J E. Environmental Biology for Engineers and Scientists[M]. Hoboken :John Wiley & Sons, Inc., 2006.)

A_2 is a mineral horizon somewhat darkened by the presence of organic matter; A_3 is a transition zone between the A and B horizons. Letters are then added in subscript to denote particular characteristics: ca means the soil contains calcium and magnesium carbonates; g (for "gleying") means the soil is poorly aerated and frequently waterlogged; m means the soil is strongly cemented together, like a soft rock.

Soils vary according to the rock from which they are derived. This affects the size of their mineral grains, ranging from coarse sand (600~2000 μm) to silt (2~60 μm) and clay (less than 2 μm), and their chemical characteristics. Soils derived from granites, for example, develop slowly, are usually sandy, and contain relatively few plant nutrients; those developed from limestones are usually fine-grained and relatively rich in plant nutrients.

Once formed, soils begin to age. The rate at which they do so depends principally on the climate and vegetation. Desert soils age slowly, and so do those in polar regions, but in the humid tropics soils age much more quickly as luxuriant plant growth extracts nutrients and returns them for decomposition into soluble forms, which are leached rapidly by the abundant water. It is possible, therefore, to describe as "young", "mature", or "ancient" soils that may have been in existence for the same length of time.

We obtain our food from soil, we erect buildings of varying weight upon it, and we use clay taken from it as construction material that may or may not be fired to make bricks. Clearly it is of great importance to us and if we are to use it the more we know about it the better. It is so variable that we cannot be satisfied in calling it simply "the" soil. It must be classified.

There have been many attempts at soil classification, the first in classical times, but it was not until the latter part of the last century that a school of Russian scientists at St Petersburg, led by Vasily Vasilievich Dokuchaev (1840~1903), proposed a theory of pedogenesis on which a formal classification could be based. It is because of this Russian origin that many soil types have Russian names, such as "podzol" and "chernozem". The Russian work laid the foundation for what is now known as "soil taxonomy", but work has continued ever since.

The system most widely used at present was devised by the US Department of Agriculture. This divides all soils into 11 orders (www.*explorer.it/aip/keytax/content.html*). The orders are divided further into sub-orders, great groups, families, and soil series.

The classification may be powerful, but there are attractions in calling Mollisols "prairie soils" (or chernozems), Histosols "peat" or "muck", and, given the widespread environmentalist concern over the degradation of some tropical soils, calling Oxisols "lateritic soils", which are the names by which they used to be known. There is also a Canadian classification system that divides soils into two orders, Brunisolic comprising 4 Great Groups and Chernozemic with 3 Great Groups and a total of 42 Subgroups.

Source: Allaby M. Basics of Environmental Science, 2nd Edition[M].The Taylor & Francis e-Library, 2002.

Words and Phrases

exfoliation [eks͵fəuli'eiʃən]   *n.* 剥落；剥落物；表皮脱落
[例句] The results show that intercrystalline corrosion and exfoliation corrosion sensibility of 7A52 aluminum alloy decrease gradually with the increase of aging temperature and time.

abrasion [ə'breɪʒn]   *n.* 磨损；磨耗；擦伤
[例句] Give me a bandage for an abrasion.

drainage ['dreɪnɪdʒ]   *n.* 排水；排水系统；污水；排水面积
[例句] Line the pots with pebbles to ensure good drainage.

orthoclase ['ɔ:θəkleɪs]   *n.* 正长石
[例句] Feldspar can be divided into two large groups, and the potassium feldspar includes orthoclase and microcline.

alga ['ælgə]   *n.* 藻类；海藻
[例句] There's also an alga that creates reddish "watermelon snow" — a phenomenon first described by Aristotle.

cyanobacterium [saɪənəubæk'tɪərɪəm]   *n.* [生] 藻青菌
[例句] Effects of solar ultraviolet radiation on cyanobacterium anabaena have been extensively studied.

decay [dɪ'keɪ]   *vi.* 衰退，[核] 衰减；腐烂，腐朽；*n.* 衰退，[核] 衰减；腐烂，腐朽；*vt.* 使腐烂，使腐败；使衰退，使衰落
[例句] The bodies buried in the fine ash slowly decayed.

anchorage ['æŋkərɪdʒ]   *n.* 锚地；下锚；停泊处
[例句] The ship remained in anchorage for a month.

decomposition [͵di:kɒmpə'zɪʃn]   *n.* 分解，腐烂；变质
[例句] Is this different from functional decomposition?

pedogenesis [͵pi:dəʊ'dʒenɪsɪs]   *n.* [医] 幼体生殖；[地] 成土作用；土壤发生
[例句] As, Cr and Ni are elements controlled by pedogenesis.

bedrock ['bedrɒk]   *n.* 基岩；基本原理
[例句] Bedrock is the solid rock in the ground which supports all the soil above it.

Questions

1. What causes the rock to break?
2. What is the main source of the carbon dioxide engaged in subsurface

carbonation?

3. How did vegetation become part of the developing soil?

4. What are soil horizons?

Part C   Dialogue

Dialogue 1

This is the VOA Special English Agriculture Report.

Studies show farmland in Africa is often lacking in important nutrients. But researchers say a combination of farming methods may help.

Since the world food crisis several years ago, researchers have directed more of their attention to small farms. Most farms in areas south of the Sahara Desert are only about one or two hectares. One of the goals is to increase production without necessarily clearing more land to grow additional crops.

American researchers say that can happen with greater use of an agricultural system called perenniation. It mixes food crops with trees and perennial plants – those that return year after year.

Soil scientist John Reganold is with Washington State University.

John Reganold: One of the major problems (is that the) soils are fairly poor in most of the regions. So how do you grow food on poor soils? There have to be food production systems that can build the soil and improve the yield.

Mr. Reganold says poor soil may have resulted from years of weathering that washed away many nutrients. He says some farmers may have done more harm than good.

John Reganold: They have been actually using farming practices where they're not putting in organic matter. They're not putting in fertilizers. They can't afford those things. And it just runs the soil down. So they're worsening the situation.

He estimates that up to two billion dollars worth of nitrogen, phosphorous and potassium is lost from African soil each year.

The scientist says the word perenniation defines three systems that are already used in Africa. The oldest of the three is called evergreen agriculture. This is where farmers plant trees with their crops. John Reganold says farmers in Africa have been doing this for sixty years, but it seems to be growing in popularity.

The method is gaining widespread use in countries such as Niger, Burkina Faso, Malawi and Zambia. The trees are planted among maize, millet or sorghum crops. They not only add nitrogen to the soil through their roots, but also through their leaves when they fall off and break down. At other times of the year, the trees can protect plants from strong sunlight.

John Reganold says he knows of one woman who has had great success with perenniation.

John Reganold: She's a grandmother in her fifties. Her name is Rhoda Mang'yana and she started using this system about twenty years ago. And her yields initially were about a ton of maize. Now with a good year she gets four tons per hectare. Four times what she was getting.

Mr. Reganold was one of three researchers who wrote a report about perenniation. It was published in the journal Nature.

And that's the VOA Special English Agriculture Report. I'm Steve Ember.

Source: http://www.tingclass.net/show-8385-263157-1.html

Dialogue 2

The report is issued jointly by China's Ministry of Environmental Protection and the Ministry of Land and Resources. It is based on a 9-year survey starting since 2005, on about two thirds of the land across the Chinese mainland.

Gao Shengda is secretary of the Industry Alliance for Environmental Rehabilitation. He says the main pollution source is human industrial and agricultural activities.

Gao Shengda: The rapid economic growth of China over the past few decades has taken its toll on our environment. Industrial waste contaminates farmland around factories and mines. Chemical plants in the suburbs also produce organic and inorganic pollutants that are hazardous to the land. Therefore, we call for stronger supervision and liability-pursuing mechanisms that focus on the disposal of waste produced by industries known for producing heavy metal waste as well as the abuse of chemical products used during agricultural production.

According to the report, irrigation by polluted water, the improper use of fertilizers and pesticides and the development of livestock breeding also cause pollution to farming land.

Gao says soil rehabilitation is a time-consuming and costly process.

Gao Shengda: Pollution is hard to eliminate because the self-purification capacity of soil and underground water is weak. China has focused on microbe and plant-restoring methods, while overseas countries use physical and chemical methods to restore the soil. It's easy for us to pollute soil, but it will cost us ten times or even one hundred times as much investment to purify it.

Facing with the alarming number, the government is taking countermeasures to better protect the soil environment and curb pollution.

The environment ministry is rushing to map out an anti-land pollution action plan in cooperation with other related departments.

Wu Xiaoqing is the deputy minister of the ministry.

Wu Xiaoqing: We are drafting an action plan on soil protection and pollution control. The plan will be based on notions of ensuring agricultural products' safety and providing a healthy living environment. That means we'll have to achieve the goal by enhancing the land quality of arable and construction sites.

Meanwhile, we'll strengthen laws which curb the soil pollution and better protect environmental safety. Moreover, regulation on pollution monitoring, control and remediation will also be enhanced.

Revising the environmental protection law, which took effect in 1989, has been deemed vital to curbing pollution.

For CRI, this is Alexander Aucott.

Source: http://www.tingvoa.com/html/20140419/China-Alerted-to-Serious-Soil-Pollution.html

Part D Supplementary Vocabulary

Soil Erosion and Soil Pollution

accelerated erosion [æk'seləreɪtɪd ɪ'rəʊʒən] 加速侵蚀
affluent ['æfluənt] *adj.* 富裕的；丰富的；流畅的；*n.* 支流；富人
agriculture measure ['ægrɪkʌltʃə(r) 'meʒə(r)] 耕作措施
agro-forestry ['ɑːgrəʊf 'ɒrəstri] 复合农林业
alluvial [ə'luːviəl] *adj.* (河流、洪水) 冲积的，淤积的；*n.* 冲积土，冲积层
ameliorate [ə'miːliəreɪt] *vt.* 改善；减轻(痛苦等)；改良；*v.* 变得更好

agricultural practice 农业技术措施
ancient erosion [ˌeɪnʃənt ɪ'rəʊʒn] 古代侵蚀
archaeological [ˌɑːkiə'lɒdʒɪkl] *adj.* [古] 考古学的；[古] 考古学上的
area suitable to control 宜治理面积
barren ['bærən] *adj.* 贫瘠的；不生育的；无益的；沉闷无趣的；空洞的；*n.* 荒地
basin ['beɪsn] *n.* 水池；流域；盆地；盆
basin lag ['beɪsn læg] 流域滞留
basin order ['beɪsn 'ɔːdə(r)] 流域等级

basin perimeter ['beɪsn pə'rɪmɪtə(r)] 流域范围

canopy density 郁闭度

capital farmland 基本农田

cash forest [kæʃ 'fɒrɪst] 经济林

catchment ['kætʃmənt] *n.* 集水，集水处（水库或集水盆地）

cementation [ˌsiːmen'teɪʃn] *n.* 黏结；水泥接合；渗碳处理

channel migration 河道迁移

check dam [tʃek dæm] 谷坊

check dam for farmland forming 淤地坝

cohesion [kəʊ'hiːʒn] *n.* 凝聚；结合；[力] 内聚力

colluvial [kə'luːvɪəl] *adj.* 塌积的

compaction [kəm'pækʃən] *n.* 压紧；精简；密封；凝结

comprehensive [ˌkɒmprɪ'hensɪv] *adj.* 综合的；广泛的；有理解力的；*n.* 综合学校

comprehensive control of soil erosion and water loss 水土保持综合治理

concentration [ˌkɒnsen'treɪʃən] *n.* 浓度；集中；浓缩；专心；集合

confluence ['kɒnfluəns] *n.* [术]（河流的）汇合处；[正]（事物的）汇合

contour tillage ['kɒntʊə(r) 'tɪlɪdʒ] 等高耕作

covering cultivation ['kʌvərɪŋ ˌkʌltɪ'veɪʃn] 覆盖种植

cretaceous [krɪ'teɪʃəs] *adj.* 白垩的，白垩纪的；*n.* 白垩纪，白垩系

dam system [dæm'sɪstəm] 坝系

critical wind velocity 临界风速，失稳临界风速

current ['kʌrənt] *adj.* 现在的；流通的，通用的；最近的；草写的；*n.*（水、气、电）流；趋势；涌流

crown density 授密度，[林] 郁闭度（树冠）

debris ['debriː] *n.* 碎片，残骸

debris flow ['debriːfləʊ] 泥石流

debris slide ['dəbriːslaɪd] 泻溜

delivery ratio [dɪ'lɪvəri 'reɪʃɪəʊ] 泥沙输移比

density of plantation 造林密度

desertification [dɪˌzɜːtɪfɪ'keɪʃn] *n.* 荒漠化

detention [dɪ'tenʃn] *n.* 拘留；延迟；挽留

diversion [daɪ'vɜːʃn] *n.* 转移；消遣；分散注意力

downstream [ˌdaʊn'striːm] *adv.* 在下游地；顺流地；*adj.* 在下游方向的

drainage basin ['dreɪnɪdʒ 'beɪsn] 流域盆地

drainage divide ['dreɪnɪdʒ dɪ'vaɪd] 流域分界线

dynamic analysis [daɪ'næmɪk ə'næləsɪs] 动态分析

ecological benefits of soil and water conservation 水土保持生态效益

ecological remediation [ˌiːkə'lɒdʒɪkl rɪˌmiː'dɪeɪʃn] 生态修复

engineering measure [ˌendʒɪ'nɪərɪŋ 'meʒə(r)] 工程措施

enmesh [ɪn'meʃ] *vt.* 使绊住；使陷入

erosion base [ɪ'rəʊʒn beɪs] 侵蚀基准面

erosion modulus [ɪ'rəʊʒn 'mɒdjʊləs]

侵蚀模数
exudate ['eksə,deɪt]　*n*. 分泌液；流出物
fallow ['fæləʊ]　*adj*. 犁过而未播种的，（指耕地）未经耕作的；*n*. 休闲地，休耕地
farm drainage ['dreɪnɪdʒ]　*n*. 农田排水
farmland formed in silt storage dam 坝地
firewood forest 薪炭林
fish-scale pit 鱼鳞坑
flood and drought disaster 水旱灾害
freeze-thaw erosion ['fri:zθɔ: ɪ'rəʊʒn] 冻融侵蚀
geomorphic [,dʒi:ə'mɔ:fɪk] *adj*. 地球形状的；地貌的
gley [gleɪ]　*n*. [土壤] 潜育土
grassland degradation ['gra:slænd ,degrə'deɪʃn] 草场退化
gravitational erosion [,grævɪ'teɪʃənl ɪ'rəʊʒn] 重力侵蚀
gully density ['gʌli 'densəti] 沟壑密度
gully erosion ['gʌli ɪ'rəʊʒn] 沟蚀
headward erosion ['hedwəd ɪ'rəʊʒn] 溯源侵蚀
herbaceous [hɜ:'beɪʃəs]　*adj*. 草本的；绿色的；叶状的
horizontal stage [,hɒrɪ'zɒntl steɪdʒ] 水平阶
hydraulic [haɪ'drɔ:lɪk]　*adj*. 液压的；水力的；水力学的
hydraulic engineering [haɪ'drɔ:lɪk ,endʒɪ'nɪərɪŋ]　水利工程（学）
hydro-junction ['haɪdrəʊdʒ'ʌŋkʃn] *n*. 水利枢纽

hydrological [,haɪdrə'lɒdʒɪkəl] *adj*. 水文学的
hydrology [haɪ'drɒlədʒi]　*n*. 水文学，水文地理学
hydrotechnics [haɪdrə'tekniks] *n*. 水利技术；水利工程学
illuviation [ɪ,lu:vɪ'eɪʃən]　*n*. [地质] 淀积作用
incise [ɪn'saɪz]　*vt*. (在表面)雕，刻
infiltrate ['ɪnfɪltreɪt]　*vt*. 使潜入；使渗入，使浸润；*vi*. 渗入；*n*. 渗透物
infiltration capacity 渗透量，渗入量
institutional factor [,ɪnstɪ'tju:ʃənl] 体制因素
inundation [,ɪnʌn'deɪʃn]　*n*. 淹没；洪水；（洪水般的）扑来；压倒
landscape pattern ['lændskeɪp 'pætn] 景观格局
landslide ['lænd,slaɪd]　*n*. 滑坡
leaching erosion ['li:tʃɪŋ ɪ'rəʊʒn] 淋溶侵蚀
level ditch ['levl dɪtʃ]　水平沟
loess ['ləʊɪs]　*n*. 黄土
meshing and sowing seeds technology 挂网喷混植草技术
Ministry of Water Conservancy and Electric Power　水利电力部
natural erosion ['nætʃrəl ɪ'rəʊʒn]　自然侵蚀
nodum ['nəʊdəm]　*n*. 植被单位
perimeter [pə'rɪmɪtə(r)]　*n*. 周长；周界；[眼科] 视野计
non-tillage [nɒn'tɪlɪdʒ]　免耕
planning of soil and water conservation　水土保持规划

plans for river basins 流域规划
plantation [plɑːnˈteɪʃn] *n.* 栽植；大农场；*adj.* 适用于种植园或热带、亚热带国家的
plough [plaʊ] *n.* 犁；耕作
protective cover of vegetation 植被防护层
recent erosion [ˈriːsnt ɪˈrəʊʒn] 现代侵蚀
recurrence interval 脉冲周期，重复间隔
regime [reɪˈʒiːm] *n.* 体制
regionalization [ˌriːdʒənlɪˈzeɪʃən] *n.* 区域化；分成地区；按地区安排
remote sensing technique 遥感技术
remote-sensing information 遥感信息
runoff plots 径流小区
salinization [səlɪnɪˈzeɪʃən] *n.* 盐化作用，盐碱化
sand barrier [sænd ˈbærɪə(r)] 沙障
sandy desert [ˈsændi ˈdezət] 沙漠
saturate [ˈsætʃəreɪt] *vt.* 浸透，使湿透；使饱和，使充满；*adj.* 浸透的，饱的；深颜色的
saturation [ˌsætʃəˈreɪʃn] *n.* 饱和度；（达到）饱和状态；浸透；饱和剂
sediment concentration [ˈsedɪmənt ˌkɒnsenˈtreɪʃn] 含沙量
sediment delivery ratio [ˈsedɪmənt dɪˈlɪvəri ˈreɪʃɪəʊ] 泥沙输移比
sediment discharge rate [ˈsedɪmənt dɪsˈtʃɑːdʒ reɪt] 输沙率
sediment runoff [ˈsedɪmənt ˈrʌnˌɔːf] 固体径流
sedimentary [ˌsedɪˈmentri] *adj.* 沉积的，沉淀性的
silt [sɪlt] *n.* 淤泥，泥沙；*vi.* 淤积；被淤塞；*vt.* 使淤积；充塞
slacken [ˈslækən] *vi.* 松劲，懈怠；变松弛；变缓慢；*vt.* 使缓慢；使松弛
slope [sləʊp] *n.* 斜坡；斜率；倾斜；斜面；*vi.* 倾斜；有斜度；悄悄地走；潜行；*vt.* 使倾斜
slope collapse [sləʊp kəˈlæps] 崩岗
slope protection [sləʊp prəˈtekʃn] 护坡
sloping terrace [ˈsləʊpɪŋ ˈterəs] 坡式梯田
slurry [ˈslʌri] *n.* 泥浆，浆
small watershed 小流域
soil and water conservation 水土保持
soil conservation measures 土壤保持措施
soil erosion and water loss 水土流失
soil erosion intensity [sɔɪl ɪˈrəʊʒn ɪnˈtensətɪ] 土壤侵蚀强度
soil erosion rate [sɔɪl ɪˈrəʊʒn reɪt] 土壤侵蚀速率
soil flow [sɔɪl fləʊ] 流土
soil maturation [sɔɪl ˌmætʃuˈreɪʃn] 土壤熟化
surface erosion 面蚀
terrace [ˈterəs] *n.* 梯田
tributary [ˈtrɪbjətri] *n.* 支流；*adj.* 支流的；辅助的
valley environment [ˈvæli ɪnˈvaɪrənmənt] 流域环境
valley harnessing 流域治理
vegetable layer 植被层
vegetable measures [ˈvedʒtəbl ˈmeʒə(r)]

植物措施
water conservancy facilities 水利设施
water erosion 水力侵蚀
watershed ['wɔ:təʃed] *n.* 流域；分水岭；集水区；转折点；*adj.* 标志转折点的

wave erosion [weɪv ɪ'rəʊʒn] 波浪侵蚀
wind erosion [wɪnd ɪ'rəʊʒn] 风力侵蚀
zone of vegetation 植被带

Part E   Supplementary Knowledge

Supplementary Knowledge 1

Action Plan for Prevention and Control of Soil Pollution

Not official translation and for reference only

Action Plan for Prevention and Control of Soil Pollution（No. 31 [2016] of the State Council）

Contents

1. Conduct Survey on Soil Pollution, and Get a Better Understanding of the Baseline of Soil Environment Quality

2. Promote Legislation of Soil Pollution Control; Establish Sound Regulation and Standard System

3. Manage Agricultural Land by Categories to Ensure a Safe Environment for Agricultural Production

4. Implement the Access Management for Construction Land; Guard Against the Risk of Livelihood Environment

5. Strengthen the Protection of Soil that are not Polluted; Strictly Control the Pollution of Unpolluted Soil

6. Strengthen the Regulation on Pollution Sources and Emphasize the Soil Pollution Prevention and Control

7. Conduct Pollution Treatment and Remediation; Improve Regional Soil Environment Quality

8. Increase Efforts on Technological Research and Development; Advance the Progress of Environmental Protection Industries

9. Give Full Play to the Dominant Role of Government and Develop a Soil Environmental Governance System

10. Strengthen Objective Assessment and Accountability

土壤污染防治行动计划

一、开展土壤污染调查,掌握土壤环境质量状况
二、推进土壤污染防治立法,建立健全法规标准体系
三、实施农用地分类管理,保障农业生产环境安全
四、实施建设用地准入管理,防范人居环境风险
五、强化未污染土壤保护,严控新增土壤污染
六、加强污染源监管,做好土壤污染预防工作
七、开展污染治理与修复,改善区域土壤环境质量
八、加大科技研发力度,推动环境保护产业发展
九、发挥政府主导作用,构建土壤环境治理体系
十、加强目标考核,严格责任追究

Supplementary Knowledge 2

如何做学术海报

1. 功能

学术海报促进科学思想的快速传播。海报是研究工作的视觉化展示,也可以用作小团体的简洁明了的传播工具。从与受众建立某种关系的角度来看,如果展示得当,它会比演讲更有效,因为海报这种展示方式允许你和对你的研究感兴趣的人一对一地互动。它不仅能吸引观众并激发好奇心,也是研究工作的宣传和总结,并且在你不在场的情况下也能被观众看到。尽管海报看起来应该专业一些,但是,实际的海报展示环节通常都不太正式并具有交互性;而在报告会上,人们反而不倾向于提问和回答问题。除了培养娴熟的科技写作技能来表达你的研究成果之外,掌握良好的展示技巧可以说是传播科学中最有效也最快捷的方法。

2. 内容

海报指南1:围绕你的研究问题设计海报
包括:题目、摘要、引言、材料与方法、结果、结论、(参考文献、致谢)
海报指南2:只关注每个章节的要领
海报指南3:用视觉化方式展示你的研究工作——通过图示说明问题

海报主要是利用文字来支持图像的一种视觉展示方式。大体来说,海报遵循标准科技论文的结构,它们包含一篇研究论文里除了"讨论"以外的所有章节。海报的内容包括:题目、摘要、引言、材料与方法、结果和结论,参考文献和致谢则是可选的。

海报这些组成部分与论文章节的主要不同之处在于海报只关注每个章节的要领，简要并视觉化地展示这些章节的内容。海报不是对论文原封不动的照搬，过于详细的方法描述或表格里冗长的数据都是不必要的。它可以用作在会议期间或会后对所展示内容感兴趣的人私下讨论的材料或讲义材料。

有时候，把海报的每一部分看作是展示给观众的一张幻灯片会对海报设计有很大帮助。像展示幻灯片一样，在整个海报中，尽量使用最少的文字，比如使用关键词。不要照搬期刊论文、会议论文摘要或其他稿件的一部分到海报展板上，而要将科学思想精心设计成引人入胜的形式展示给观众。

- 围绕研究问题来设计海报。在你的海报展示时间里，你可以围绕你的研究主旨与他人展开讨论。
- 提供明确的、可以带回家的信息。针对有教育背景的观众，简要地总结研究启示和结论。根据会议类型的不同，你可能还需要针对你的研究领域里的专家来调整海报的设计。
- 对于对研究有贡献的人应该表示感谢。如有需要，将"致谢"用较小字号(14~18 磅)写入海报，以表示对贡献者和资助机构的感谢。

要记住展示的清晰度在于对信息恰当的安排，海报的美观度通常在于简洁的设计。原因在于海报是一种视觉展示，你需要用视觉化的方式展出你的研究工作。使用示意图、箭头和其他方法来引导观众的注意力，而不是全部只用文字来解释。

3. 结构

海报指南 4：使海报每部分的篇幅和空间不尽相同以增加视觉效果

总体结构

评判学术海报的标准不仅包括内容也包括它的设计编排。因此，有效的设计非常重要。当你需要展示海报时，要做的第一件事就是搞清楚你可以使用多大空间，这决定了你可以展示多少细节。与文稿不同，海报可以采用各种各样的布局。只要你给出足够的空白，有逻辑性地排列栏目，并给读者提供清晰的线索引导他们"游历"你的海报内容，你就可以创造性地设计自己的海报。但是，考虑一下各部分的最佳位置。在一幅海报里，中央部分是图示的最佳位置，左上角和右下角是文字的最佳位置。因此，你的最重要的图示和结果应该放在海报中央，最重要的文字部分(摘要和结论)应该放在相对应的文字最佳位置，分别是左上角和右下角。为清楚起见，总结如下：

- 将信息按照易读的顺序编排。你或许喜欢使用数字编号来帮助排序或利用标题来显示信息的走向。
- 利用栏目来编排材料——大多数海报允许有 3~4 个栏目。
- 摘要放在左上角，结论放在右侧偏下的位置。这样，就考虑将不重要的文字部分(参考文献和致谢)放在海报的最下方位置。

- 使海报每部分的篇幅和空间不尽相同以增加视觉效果,但是不要差别太大。
- 自始至终保持一致的风格。不一致的风格给人不和谐的印象,分散读者的注意力且有可能影响信息的流畅性。确保每页海报的标题都出现在同一位置。

海报背景和颜色

关于海报背景的设计:

正确

- 使用彩色背景来统一海报。采用何种背景颜色取决于你。你也可以使用第二种背景颜色把海报的每部分框起来。如有需要,可将图先固定在一个不同于其他颜色的单色薄板上或用带状颜色勾勒出图的轮廓。
- 使用柔和的颜色作背景。这种颜色适合长时间观看并能最好地衬托文字、图表和照片。
- 对于深色图片,使用浅色背景;对于浅色图片,使用深色背景。使用非彩色的(灰色)背景来强调图片里的颜色,使用白色背景来减弱彩色图片对视觉的冲击。
- 在使用颜色上保持一致的模式。否则,观众会花费很多时间来看懂你的模式而削弱了对海报的注意力。
- 同时,还需要考虑到那些在颜色辨认上有困难的人,尤其是在设计图表的时候。最普遍的色弱或色盲现象之一是不能辨别红色和绿色。

文本格式

准备一张海报与准备一篇文章大不相同。为海报展示所准备的文字是被提炼过的非常简洁的语言。要知道人们容易被图表美观、标题正确、文字简短的海报所吸引。

推荐字体和字号

- 使用无衬线字体(san serif),如 Arial,而不是 Times New Roman。无衬线字体比衬线字体更容易阅读。
- 字号大小需要使读者在 1.8 米以外能看清楚。
- 根据文字的重要性调整字号大小。
  - 标题: 9 磅,黑体
  - 副标题: 72 磅
  - 章节标题(引言等): 32~36 磅
  - 其他文字: 理论上 22~28 磅,黑体,1.5~2 倍行距
- 注意图中的字号大小——同样也需要使用大号字体。
- 使用斜体进行强调。

正确

- 使用有趣的标题，大号字体显示。
- 只使用少量文字。
- 为海报安排好结构，引导读者阅读。
- 使用项目符号和编号来分割文字，使文字更易读。
- 所有文字使用两倍行距，左对齐；左对齐而右侧参差不齐的文字是最易读的。
- 使用主动语态。
- 删除所有冗余的参考文献和填充词。
- 在最终印刷之前仔细检查拼写和校对文字。

不正确

- 使用过多文字和信息。
- 使字号太小。
- 随意排列展板。
- 将会议摘要用作海报上的文字。
- 改变海报上的字体(要自始至终使用同一种字体)。
- 使用下划线和惊叹号，因为其作用不大。

4. 组成部分

海报指南 5：精简文字，用图表来代替文字

因为海报本质上是一种视觉媒介，所以你应该使用最少的文字和尽可能多的图表。使用短句、简单的词语，并用项目符号来示例各个分项。不要使段落过长，要使用短小、简单的陈述。同样需要考虑基本的写作原则，例如：使用主动语态，避免专业术语和冗余。

标题

标题需要阐明主题、方法和体系(生物体)。它需要吸引读者的注意力，应不长于两行，并至少 5 厘米高。

摘要

有些海报包含摘要，有些不包含。如果摘要已经发表，就没有必要在海报中重复了。如果要将摘要放入海报中，尽量精简文字。参照摘要的写作指南，确保包含了所有重要信息。摘要应限制在 50~100 个词。

引言

在使用最少背景知识和定义的情况下，让观众对你的课题或问题感兴趣。快速表明你的研究在已有的文献和研究背景中的位置，清晰表述你的研究问题，并

简要描述大概的实验方法并证明其有效性。将背景知识减少到最少，如果需要，你可以现场补充细节。引言部分应该简短一些(不超过 200 个词)。

材料与方法

简要描述实验设备和方法。描述不要像论文那样详细，尽可能地使用图表来说明实验设计。如果需要，使用流程图来总结实验过程。将照片、绘制图和相关参考文献放入其中——文中只出现姓名和日期即可。同样，篇幅限制在 200 个词以内。

结果

描述最重要的和总的研究结果。对于基础研究类的海报来说，这部分会是最占篇幅的一部分。你的大部分研究发现，如果不是全部的话，都应该以图表的形式展现。给出数据分析，这会逐渐引出主要研究发现。提供引人入胜、有独立性、并能解释研究发现的图注说明，尤其是在结果部分没有其他文字描述的情况下。给表格也配以表注说明。确保结果和结论保持一致的叙述顺序。有时候，结果和结论通常以分项列举的形式展示在一张展板上。

结论

海报的最后一部分通常是结论部分。讨论部分留待给观众讲述并在论文中发表。在结论中，阐述研究结果的重要性，并相应地进行总结。保持主要词语在海报中的一致性，利用项目符号、箭头、斜体或彩色文字来强调重点。因为海报只表现整个研究的一部分，所以结论通常是非常简短的。集中笔墨在你的主要发现和对它们的阐释上，不要列举所有的研究发现。在结论或总结中，只需提到 2~4 个重点。而将其写成分项的形式会比一整段文字的形式在视觉上更加吸引读者。作为结论的一部分，可以考虑绘制一个模型或一幅图片以突出论文中提出的假说、模型或机理。

有时候，结论之后还有一个短小的章节题为"未来的研究方向"。这部分用来简要叙述你计划怎样扩展目前的研究。

参考文献

如果海报引用了他人的技术方法，需要提供参考文献，但是要尽量简短。通常情况下，提供姓名、日期和刊登期刊就足够了。海报上文内的引用只提供姓名和年份即可，尽可能不要超过 5 个引用。

致谢

在致谢部分感谢为研究做出贡献的个人(例如：提供设备，提供统计方面的建议，实验方面的帮助，对海报前几个版本的意见)，并写明提供基金的资助机构。同时，还需要公开任何可能的利益冲突和承诺冲突。

5. 照片、图片和表格

海报指南 6：图表要简单易懂
海报指南 7：图表要有吸引力

视觉辅助手段的大致内容和形式

一张海报的图表的清晰度直接关系着这张海报的成败。海报中应该尽量使用不言自明的图表。在会议召开之前就早早准备好各种图表可以使你有时间检查、替换或改进它们。与文字一样,图表材料的大小也应该使读者在离海报至少 1.8 米的地方可以看清楚。

图和表

所有图表都应该不言自明,要简单并少用文字。它们要吸引读者但又不失专业性,单个图片应该与其所属的同一系列图片保持一致的风格(颜色、字体和强调符号一致)。大多数观众会先看海报中的图表。这些图表不仅要在视觉上吸引读者,而且应该比会议论文等出版物里的图表简单。有时候可能会需要绘制流程图这样的图解,而这一般不出现在论文中。

以下指南有助于设计出尽可能好的图表:
- 用图表来说明问题
- 给每个图表都加上标题
- 给每个图表都加上图注和表注
- 以特定的名字来命名表格而不是数字或字母
- 用坐标图来代替表格
- 如果使用表格,务必要简单易懂
- 使用反差和颜色来作强调,尤其是区别图表中不同组的数据
- 使图表里的注解和线条比平常大一些,使用容易区分的符号
- 水平书写所有文字
- 解释每一个变量及其意义
- 所有图中使用统一标尺;所有坐标图中使用统一坐标轴
- 公式中字体的大小要使读者在 1.8 米外可以看清楚
- 删除图表中所有不必要的信息,使用较大字体和不同颜色直接标明数据线
- 条形图中的条形应不超过 6 个,线形图中的线应不超过 4 个
- 包括标题和列标题在内,表格不要超过 4 列、8 行
- 避免使用不标准的颜色、字体、图表和缩写
- 柱形图中避免使用图案和开放的柱形
- 删除不必要或不重要的公式

6. 准备一张海报

如果你计划在会议上展示一张海报，通常需要先撰写一份摘要，会议组委会会评审这份摘要。如果你应邀参加会议，组委会会通知你应该准备口头报告还是海报展示。摘要会对参会者公开并通常会发表。你还可能被邀请撰写一篇文章，这篇文章会被发表在会议论文集里，摘要（和海报）可以展示初步的研究成果或接近出版水平的成果。摘要需要切合会议主题，根据摘要设计出来的海报要有趣味性，能够吸引观众。

检查，检查，再检查

海报指南 8：修改的时候毫不留情

打好海报每部分的草稿后，要仔细检查有没有错误，易读性如何，风格是否一致。如果需要，可以尝试不同的布局。请朋友、同事和你的导师等其他人从布局和内容上给你建议。修改的时候要毫不留情。

7. 海报展示

海报指南 9：在海报展示环节，在海报一侧回答观众提出的问题并向他们介绍你的研究工作

提早到达展示场地。自带一套悬挂海报用的工具，除非你非常确定会议组委会将提供这些用具。把你的海报挂得整齐好看，不要侵占旁边海报的位置。可以考虑给读者提供讲义传单或海报的缩小版。这种讲义传单可以在设计海报的时候用版面设计程序制作出来。把传单、名片和海报缩小版放在海报附近——放置在桌上或悬挂在海报附近以使读者经过的时候可以获取。如果海报要展示很长时间，不要忘记隔段时间就补充新的资料。

虽然你展示和提供的资料应该传达了你的核心信息，但是在海报展示期间，最好守在海报一侧以便与观众交流讨论。另外，要准备一个 5~10 分钟的演讲，突出海报的重点。作为一个展示者，你的任务还有回答读者提出的问题，进一步提供细节信息和使读者相信你的研究工作是出色并且有意义的。在实际做展示的时候，把重点放在图表上。把海报用作一种视觉辅助手段，不要逐字去读海报的内容。告诉观众你的研究问题的背景、它的重要性、你的研究目标、达到目标的方法，还有数据以及它们的意义。

Reference: Angelika H. Hofmann. 科技写作与交流-期刊论文、基金申请书及会议讲演[M]. 任胜利，莫京，安瑞，等，译. 北京：科学出版社，2012.

Unit 7 Solid Waste

Part A Intensive Reading

Solid Wastes

Solid wastes are defined as unwanted materials that are no longer of values to their owners (albeit they may be still of values to other parties). They are generated from various human and animal activities. Each year, billions of tones of solid wastes are generated which are in need of proper treatment. Improper management of solid wastes has direct adverse effects on human health and the environment (e.g., polluting rivers and groundwater sources and generating foul air).

1. Sources and Composition

The information on waste composition and generation rate is important for proper planning, treatment and disposal of solid wastes, which depends on their sources and types. Some solid wastes (e.g., industrial, agricultural and mining) are managed by the waste producers. In this unit, the focus is on the municipal solid wastes (MSW) which include residential, commercial, institutional, construction and demolition.

Another way to classify wastes is to divide them into: biodegradable (spoiled food, green waste), recyclable (paper, glass bottles, cans), inert (construction and demolition waste), electrical and electronic (TVs, computers, screens, etc.), composite wastes (waste clothing, Tetra Packs, waste plastics such as toys), hazardous (paints, chemicals, fluorescent tubes, spray cans, fertiliser), toxic (pesticide, herbicides, fungicides), and medical (pharmaceutical drugs).

As a rough rule of thumb, the residential waste generation in industrial countries is about 1 kg per person per day and the total MSW is about 1 tonne per person per year. The composition and quantities of solid wastes change with time and vary between different countries (and regions).

There are several ways to collect solid waste information: 1) literature review (the information is collected from the past data in the literature); 2) input-output analysis (this is based on the consumption data to estimate wastes); 3) sampling survey (collection of actual data by sampling survey. It needs one or two years of survey to smooth out seasonal changes. A large number of samples are needed to reduce the uncertainty).

2. Waste Properties

Solid waste management options depend on physical and chemical characteristics of the waste. Waste generation rates are expressed in mass unit (e.g., kg/m^3) instead of volume unit. This is because waste densities vary greatly among the wastes with different compaction at different stages in the waste system (collection, storage and disposal). For example, the density for mixed loose MSW wastes is 90~180 kg/m^3, 300~420 kg/m^3 in compactor truck, 480~770 kg/m^3 in initial landfill and 700~1100 kg/m^3 in overburdened landfill.

The information on waste moisture content is used to derive dry mass of the received waste. In addition, for energy recovery design, it is important to know the energy content in wastes. Plastic and paper have high energy content and low moisture content. In contrast, garden wastes are less productive in energy release due to their high moisture content.

1) Solid Waste System

a. Storage, Collection and Transport

This part of the waste system typically accounts for 40 to 80 percent of the total cost. For residential wastes, the most common collection method is kerbside. The wastes are usually segregated into different containers according to waste types (food, garden waste, recyclables) as shown in Figure 1. Different trucks may be used to collect those containers. Commercial and institutional wastes generally need larger storage containers and use a separate system for waste collection. Drop-off stations for recyclables are also a valuable part of a waste collection system where people can drop their wastes. They are more efficient for trucks to pick up at a few key points instead of travelling along all the streets to pick up individual household containers. For old cities with narrow roads or high density urban areas, the Automated Vacuum Collection (AVAC) system is able to transport waste at high speed through underground tunnels to a collection station where it is compacted and sealed in containers. When the container is full, it is transported away and emptied by trucks. The system cannot carry large items such as furniture, construction wastes or glass which may break apart and sends shards flying through the tubes. For large cities and towns, transfer stations are often used as places where local waste collection vehicles will deposit their waste cargo prior to loading into larger vehicles. These larger vehicles (specialised with higher efficiency for heavy load) will transport the waste to the end point of disposal in an incinerator, landfill, hazardous waste facility, or for recycling.

Figure 1 Waste containers for different waste types in Bristol, England
(Source: http://www.bristol.gov.uk/nav/recycling-and-waste)

b. Recycling

Recycling is to convert wastes into new products. It can reduce the consumption of fresh raw materials, energy usage, air pollution (from incineration) and water pollution (from landfill). Recyclable materials include glass, paper, metal, plastic, textiles, and electronics. Plastics recycling is very challenging due to a diverse range of plastic types. To help with plastics recycling, an international resin code is marked on most plastic products with numbers 1 to 7 (1: PET for soft drink and water bottles; 2: HDPE for milk bottles; 3: PVC for shampoo bottles, windows, and piping; 4: LDPE for shopping bags, squeezable bottles; 5: PP for syrup bottles and straws; 6: PS for egg cartons, compact disc cases; 7: all other plastics). Types 1 and 2 are the most commonly recycled.

Wastepaper is of higher value when the paper fibres are longer and less contaminated by impurities (e.g., office papers are of higher value than glossy magazines). Aluminium is of high value due to the high energy required to process aluminium ore. Ferrous metals (iron, steel) have been traditionally recovered by scrap metal processors. Although the system to process waste glass into new glass is well developed, the large cost of transportation to a glass smelter could make it impractical. Some debris from construction and demolition could be reused (such as tiles, bricks)/recycled (metal, stone, concrete). Sometimes there is a limit on the times of recycling circles (e.g., materials such as paper pulp can only be recycled a few times before material degradation prevents further recycling).

Separation of recyclable wastes can be done either by machines or human. Magnets can separate ferrous metals. Papers and plastics may be separated by their differences in density and sizes using screens, shaking tables, burst of air and rotating sieves. Glass materials with different colours and aluminium materials can also be separated by machines. In many situations, people are still needed to aid the separation of waste materials (e.g., to pick up specific wastes from a conveyor). Materials to be recycled are either brought to a collection centre or picked up from the kerbside, then sorted, cleaned, and reprocessed into new materials bound for manufacturing. Successful recycling requires careful consideration of the processing capacity and the markets for the recycled goods (e.g., a potential demand for the recycled products). If neither of them exists, recycling is incomplete and in fact only "collection".

The cost-benefit of recycling is complex. Economically, a major benefit is the reduced landfill costs. Other benefits can be worked out by comparing the market cost of recyclable material to the cost of new raw materials. What is difficult to estimate is externalities, which are unpriced costs and benefits that accrue to individuals outside of market transactions, such as decreased air pollution and greenhouse gases from incineration, reduced hazardous waste leaching from landfills, reduced energy consumption, and reduced waste and resource consumption, etc. For example, creating a new piece of plastic may cause more pollution and be less sustainable than recycling a similar piece of plastic, but these factors will not be counted in market cost. A life cycle assessment can be used to determine the levels of externalities and decide whether the recycling may be worthwhile despite unfavourable market costs. Alternatively, legal means (such as a carbon tax) can be used to bring externalities into the market, so that the market cost of the material becomes close to the true cost.

c. Composting

Composting is a microbial process used to treat biodegradable wastes. This is similar to aerobic wastewater treatment. Composting has several purposes: 1) to reduce waste mass; 2) to reduce pollution potential; 3) to destroy pathogens; and 4) to produce compost as nutrients for gardens, landscaping and agriculture (organic farming).

This processing is suitable for garden waste, food waste, paperboard, etc. Composting organisms require four equally important things to work effectively: carbon (to produce heat by the microbial oxidation of carbon), nitrogen (to grow and reproduce more organisms), oxygen (for oxidising the carbon as part of the decomposition process) and water (to maintain activity without causing anaerobic conditions). The most efficient composting occurs with a carbon: nitrogen mix of about 30 to 1. Nearly all plant and animal materials have both carbon and nitrogen, but amounts vary widely. When one waste is not compostable on its own, its mixing with other materials may help to build a proper carbon: nitrogen mix, air porosity and pH.

Composting requires making a heap of wetted organic matter and waiting for the materials to break down into humus after a period of weeks or months. Modern composting is a multi-step, closely monitored process with measured inputs of water, air and well mixed organic materials. The decomposition process is aided by shredding the plant matter, adding water and ensuring proper aeration by regularly turning the mixture. Worms and fungi further break up the material. Aerobic bacteria manage the chemical process by converting the inputs into heat, carbon dioxide and ammonium. The ammonium is further converted by bacteria into plant-nourishing nitrites and nitrates through the process of nitrification. In addition to the traditional compost pile, various approaches have been developed to handle different composting processes, ingredients, locations, and applications for the composted product. Nowadays, industrial composting systems are increasingly being installed as a waste management alternative to landfills.

d. Incineration

Incineration (waste-to-energy, energy recovery) is a process to liberate the energy in waste by combustion. This is suitable for wastes with high energy content, low moisture content and low ash content, which include paper, plastics, textiles, rubber, leather and wood. Incineration produces two solid by-products: bottom ash (unburnt waste) and fly ash (suspended particulate matter). Both of them contain hazardous matters in need of careful management. Incinerators reduce the solid mass of the original waste by 80%~85% and the volume by 95%~96%, therefore the requirement for landfill is significantly reduced.

e. Landfill

A landfill site is a site for the disposal of solid wastes by burial. This is the oldest form of waste treatment and the most common method for waste disposal in many places of the world. In a landfill, wastes are packed into solid forms and covered to insulate them from water and air (usually everyday with layers of soil). Leachate is a liquid drained from the landfill as a contaminated wastewater. In a new landfill with oxygen still available, organic matters decompose and carbon dioxide is produced. As time goes by, the remaining oxygen is consumed and an anaerobic condition leads to the production of methane gas.

Landfill involves siting, engineering design, construction, operation, monitoring and closure. Landfills should be located with a minimum risk to the environment and society (e.g., to avoid floodplains, active geological faults, drinking-water catchments, etc.). Geographical Information System (GIS) is a useful tool in helping design engineers in site selection. The landfill gas is a source of energy to produce electricity. The amount of methane produced from the landfill is about 100 L/kg. Half-lives for methane production vary widely depending on waste compositions (from 1 to 35 years). Although landfill is covered, small quantity of water may still be able to get through to produce leachate. Concentration of leachate constituents changes with time and is usually much higher than those in untreated urban sewage. Hydraulic barriers to cover and underlie the landfill should be constructed with compacted clay and geomembranes to limit the rainfall infiltration and leachate leaking. A collection system based on gravity is used to convey leachate to a sump within the landfill and then to be pumped out to a storage location. Landfill leachate is similar to urban sewage and the collected leachate should be sent to a wastewater treatment plant by pipe or truck. A landfill requires careful management over its lifetime aided by a sound monitoring system. In recent years, some countries, such as Germany, Austria, Belgium, the Netherlands, and Switzerland, have banned the disposal of untreated waste in landfills, and only the ashes from incineration or the stabilised output of mechanical biological treatment plants may still be deposited.

2) Alternative Technologies

In addition to the aforementioned traditional technologies, many new approaches are undergoing development. They include anaerobic digestion, alcohol/ethanol production, biodrying, gasification, in-vessel composting, mechanical biological treatment, mechanical heat treatment, plasma arc waste disposal, pyrolysis, tunnel composting, and waste autoclave.

3) Solid Waste Management

Solid waste management is a complex task and requires a systems approach. The combination of different components should be considered as a whole in a holistic way to serve the need of the current and future generations. Although economy of scale means larger landfills, incinerators and composting plants are more economically efficient, they would increase transport cost and are more likely to face public opposition. Consultation with stakeholders is an important step for any successful solid waste management. Stakeholders with direct and indirect interests in solid waste management may include neighbours, local communities, wider communities, media, government agencies and various social interest groups. Good policy should be developed by analysing policy options with assessment in costs and benefits, risks and unintended effects. The priority in a waste management policy is to reduce the wastes and followed by "reuse" and "recycle" (the 3 Rs).

Source: Han D. Concise Environmental Engineering[J]. Epoca, 2012.

Words and Phrases

demolition [ˌdeməˈlɪʃn]   n. 拆除（等于 demolishment）；破坏；毁坏
kerbside [ˈkɜːbsaɪd]   n. 马路边；街边（等于 curbside）
incineration [ɪnˌsɪnəˈreɪʃn]   n. 焚化；烧成灰
contaminate [kənˈtæmɪneɪt]   vt. 污染，弄脏
impurity [ɪmˈpjʊərəti]   n. 杂质
composting [ˈkɒmpɒstɪŋ]   n. 堆肥化；堆制肥料；v. 把…做成堆肥
aerobic [eəˈrəʊbɪk]   adj. 需氧的，有氧的；有氧健身的
pathogen [ˈpæθədʒən]   n. [基医] 病原体；病原菌；[基医] 致病菌
suspended [səˈspendɪd]   adj. 暂停的，缓期的，悬浮的；v. 暂停；悬；延缓

foul air 恶臭气体
generation rate 生产率
municipal solid wastes (**MSW**) [mjʊˈnɪsɪpl]   城市固体废物
tetra packs 果汁和罐装饮料；四角塑料包；灌装机
rough rule 近似法则
sampling survey 采样调查
mass unit 质量单位
volume unit 体积单位
moisture content [ˈmɔɪstʃə(r)]   水分含量
dry mass 干重
ferrous metals [ˈferəs]   黑色金属
life cycle assessment 生命周期评估
air porosity 空气孔隙率

Questions

1. What are the principal ways to dispose solid wastes?
2. What does a successful recycling require?
3. What are the purposes of compositing?
4. What is the priority in a waste management policy?

Sentence-making

1. foul air, neighbor, combustion
2. incineration, SO_2, waste
3. solid waste, disposal, management
4. rubbish, generation, control
5. waste components, source, industry

Part B Extensive Reading

Waste Production must Peak this Century

Without drastic action, population growth and urbanization will outpace waste reduction, warn Daniel Hoornweg, Perinaz Bhada-Tata and Chris Kennedy.

Solid waste — the stuff we send down our chutes, discard at work and put on the curb every week — is a striking by-product of civilization. The average person in the United States throws away their body weight in rubbish every month. When waste management works well, we give it little thought: out of sight and, usually, quickly out of mind. Discarded materials are collected, some are recycled or composted, and most are landfilled or incinerated. But the global view is troubling.

In the past century, as the world's population has grown and become more urban and affluent, waste production has risen tenfold. By 2025 it will double again. Rubbish is being generated faster than other environmental pollutants, including greenhouse gases. Plastic clogs the world's oceans and rivers, causing flooding in developing-world cities. Solid-waste management is one of the greatest costs to municipal budgets.

The waste problem is acute in emerging cities. Landfills such as Laogang in Shang-hai, China; Sudokwon in Seoul; the now-full Jardim Gramacho in Rio de Janeiro, Brazil; and Bordo Poniente in Mexico City vie for the title of the world's largest. Each typically receives more than 10,000 tonnes of waste per day.

As city dwellers become richer, the amount of waste they produce reaches a limit. Wealthy societies tend to curb their waste. So as living standards around the world rise and urban populations stabilize, global solid-waste generation will peak.

Just when is difficult to predict. But by extending current socio-economic trends to 2100, we project that "peak waste" will not occur this century. Unless we reduce population growth and material consumption rates, the planet will have to bear an increasing waste burden.

Urban Problem

Solid waste is mostly an urban phenomenon. In rural communities there are fewer packaged products, less food waste and less manufacturing. A city resident generates twice as much waste as their rural counter-part of the same affluence. If we account for the fact that urban citizens are usually richer, they generate four times as much.

As urbanization increases, global solid-waste generation is accelerating. In 1900, the world had 220 million urban residents (13% of the population). They produced fewer than 300,000 tonnes of rubbish (such as broken household items, ash, food waste and packaging) per day. By 2000, the 2.9 billion people living in cities (49% of the world's population) were creating more than 3 million tonnes of solid waste per day. By 2025 it will be twice that — enough to fill a line of rubbish trucks 5,000 kilometres long every day.

Together, the member countries of the Organisation for Economic Co-operation and Development (OECD) are the largest waste generators, producing around 1.75 million tonnes per day. This volume is expected to increase until 2050, owing to urban population growth, and then to slowly decline, as advances in material science and technology make products smaller, lighter and more resource efficient.

Some countries generate more waste than others. Japan issues about one-third less rubbish per person than the United States, despite having roughly the same gross domestic product (GDP) per capita. This is because of higher-density living, higher prices for a larger share of imports and cultural norms. Waste quantities worldwide can also vary seasonally, by up to 30%, as horticultural and food wastes fluctuate. For example, household waste volumes double in the week after Christmas in Canada.

Waste reduction and dematerialization efforts in OECD countries are countered by trends in east Asia, particularly in China. China's solid-waste generation is expected to increase from 520,550 tonnes per day in 2005 to 1.4 million tonnes per day in 2025. East Asia is now the world's fastest growing region for waste, a distinction that is likely to shift to south Asia (mainly India) in 2025, and then to sub-Saharan Africa around 2050.

As a country becomes richer, the composition of its waste changes. With more money comes more packaging, imports, electronic waste and broken toys and appliances. The wealth of a country can readily be measured, for example, by how many mobile phones it discards. Solid waste can thus be used as a proxy for the environmental impact of urbanization. Most of a material's impact is through production and use. Less than 5% stems from waste management, which includes emissions from collection trucks, landfills and incinerators.

Peak Waste

The rate at which solid-waste generation will rise depends on expected urban population and living standards growth and human responses. In 2012, a World Bank report by D.H. and P.B.-T., *What a Waste*, estimated that global solid-waste generation would rise from more than 3.5 million tonnes per day in 2010 to more than 6 million tonnes per day in 2025. These values are relatively robust, because urban populations and per capita GDP can be well forecast for several decades.

Extending those projections to 2100 for a range of published population and GDP scenarios shows that global "peak waste" will not happen this century if current trends continue (see "When will waste peak?"). Although OECD countries will peak by 2050 and Asia–Pacific countries by 2075, waste will continue to rise in the fast-growing cities of sub-Saharan Africa. The urbanization trajectory of Africa will be the main determinant of the date and intensity of global peak waste.

Using "business-as-usual" projections, we predict that, by 2100, solid-waste generation rates will exceed 11 million tonnes per day — more than three times today's rate. With lower populations, denser, more resource-efficient cities and less consumption (along with higher affluence), the peak could come forward to 2075 and reduce in intensity by more than 25%. This would save around 2.6 million tonnes per day.

Convert and Divert

How can today's situation be improved? Much can be done locally to reduce waste. Some countries and cities are leading the way. San Francisco in California has a goal of "zero waste" (100% waste diversion by reduction and recycling) by 2020; already more than 55% of its waste is recycled or reused. The Japanese city of Kawasaki has improved its industrial processes to divert 565,000 tonnes of potential waste per year — more than all the municipal waste the city now handles. The exchange and reuse of materials connects steel, cement, chemical and paper firms into an industrial ecosystem.

North America and Europe have tried disposal fees, and found that as fees increase, waste generation decreases. Another tactic is to steer people to buy less with their increased wealth, and to spend more on experiential activities that require fewer resources.

But greater attention to consumption and improvement in waste management is needed in rapidly urbanizing regions in developing countries, especially in Africa. Through increased education, equality and targeted economic development, as in the sustainability scenario we evaluated (SSP1), the global population could stabilize below 8 billion by 2075, and urban populations shortly there-after. Such a path reflects a move towards a society with greater urban density and less overall material consumption. Also needed is a widespread application of "industrial ecology" — designing industrial and urban systems to conserve materials. This begins with studies of the urban metabolism—material and energy flows in cities.

Reducing food and horticultural waste is important—these waste components are expected to remain large. Construction and demolition also contribute a large fraction by mass to the waste stream; therefore, building strategies that maximize the use of existing materials in new construction would yield significant results.

The planet is already straining from the impacts of today's waste, and we are on a path to more than triple quantities. Through a move towards stable or declining populations, denser and better-managed cities consuming fewer resources, and greater equity and use of technology, we can bring peak waste forward and down. The environmental, economic and social benefits would be enormous.

Source: Hoornweg D, Bhadatata P, Kennedy C. Environment: waste production must peak this century[J]. Nature, 2013, 502(7473): 615-617.

Words and Phrases

chute [ʃuːt]　*n*. 降落伞；斜槽，滑道；*vt*. 用斜槽或斜道运送；*vi*. 顺斜道而下，在斜槽或滑道中滑行
[例句] The flow of fine glass beads at low velocity in an inclined chute was investigated by means of tracer particles.

horticultural [ˌhɔːtɪˈkʌltʃərəl]　*adj*. 园艺的
[例句] The mental health benefits of gardening are so strong that a field of medicine called horticultural therapy has been developed to help people who have psychiatric disorders deal with their conditions.

proxy [ˈprɒksi]　*n*. 代理服务器；代表权；代理人，代替物；委托书
[例句] Price differences are used as a proxy for differences in quality.

metabolism [məˈtæbəlɪzəm]　*n*. 新陈代谢；代谢作用
[例句] Basal metabolism is much lower for creatures in cold water.

drastic action 剧烈反应
solid-waste management 固体废物管理
food waste 餐厨垃圾
the organisation for economic co-operation and development (OECD) 经济合作与发展组织(经合组织)
Asia-Pacific countries 亚太国家
industrial ecology 工业生态学

Questions

1. Why Japan issues about one-third less rubbish per person than the United States, despite having roughly the same gross domestic product (GDP) per capita?

2. What is the rate at which solid-waste generation will rise depends on?

3. How to improve in waste management in rapidly urbanizing regions in developing countries?

Part C Dialogue

Interview 1

"The de-centralised waste management model adopted by Pune Municipal Corporation is one of the most successful in the country. This cost effective and eco-friendly model has been highly recommended by Maharashtra government," Municipal Commissioner Mahesh Pathak says.

What are the waste management programmes implemented by the Pune Municipal Corporation?

PMC has adopted an integrated system to process garbage collection in the city. Various agencies/organisations like SWACH (Solid Waste Handlers and Collectors' Society) and Mahila Sanghatana are engaged in the collection of waste from across the city. Out of the 144 municipal wards, 122 are covered under this programme. Around 1800 people work under SWACH and each rag-picker covers 250~300 households everyday. The Corporation has implemented the twin-bin system–green bin for collecting the organic waste and white for recyclable waste and charges 10 per month/house. It has also equipped the rag-pickers with hand-cart, push-cart, buckets and safety gears. They are also covered under medical insurance scheme. Out of the nine lakh properties, nearly three lakh already have bins in place.

At the primary level, the waste identified by the corporation is collected from various locations by SWACH and transported to the processing unit at Hadapsar. The organic waste is sent to bio-gas plants. PMC has installed around 15 bio-gas plants, each with a capacity to process five metric tonnes (mt) of waste and with a total production of 300 cubic metres of methane gas, amounting to 450 units of power per day. The power generated from the bio-gas plant is used for lighting the street lamps in and around the plant areas.

Besides, with citizens cooperation, certain areas are achieving zero waste. Like in the Cantonment, the residents manage their waste in most places while the rest is maintained by PMC. There are Mohalla Committees and Resident Associations working towards 100% segregation. Some wards are also doing segregation at source to achieve "zero-garbage" and are directly transporting segregated waste to the dumping site. All the processes are codified and training manuals too have been prepared. The successful implementation of various waste management schemes has resulted in the Katraj ward getting ISO certification. This model could be replicated for other municipal wards in future.

In a further move to encourage citizens to treat waste at source, PMC is providing a rebate of 5% on property tax to those engaging in vermi-composting practice. The Magarpatta city which has nearly five lakh flats generates five tonnes of waste everyday. They have an integrated waste management system and have also installed bio-gas plants with two tonnes capacity within the premises. The energy generated from these plants is used for operating garden pumps, etc. This helps save excessive power requirements of the township. The non-biodegradable waste is disposed off safely and the re-usable waste is sold in scrap. PMC has made it mandatory for townships to have a vermi-compost or a biogas plant in their premises.

Through good audio/visuals and traditional media like kirtans, bhajans and street shows, the Corporation is trying to bring awareness among people. The Maharashtra Chamber of Commerce, Janwani (a NGO working in the field of governance) and some companies/industries are also sponsoring the maintenance of certain wards.

How is the waste being treated?

Since June 2010, we have stopped open dumping. At the Uruli Devachi dumping site, the waste is being processed with the help of Hanjar Biotech. In another project, Rochem Separation India Ltd under the DBOOT will be generating around 11.5 MW electricity from waste using the pyrolysis/gasification technology. The methane gas that generates from the 25,000 mt of garbage will be collected by the corporation. We are now claiming for Certified Emission Reductions or CERs (a type of emissions unit or carbon credits) issued by the Clean Development Mechanism (CDM) body. At present, we are managing with the Uruli plant but in case of a breakdown or the ramp becoming non-functional, it is very difficult to manage the waste. About 50%~60% waste coming to this dumping ground is non-segregated garbage. The commission of the Rochem's plant will increase garbage processing capacity and the plant will be able to take two~three days stock at a time. Our aim is to tackle 50 tonnes of waste in the ward itself by creating more bio-gas facilities.

PMC collects around 125 tonnes of organic waste from commercial establishments, 90~110 mt from rag-pickers and 50~70 mt from the market. While the collection is done in two shifts, the organic waste, especially from hotels, is put into separate bio-degradable bags.

At one level PMC has made vermi-composting mandatory for townships; at another level it is spreading awareness through traditional media and good audio/visuals

The Corporation has around 200 vehicles for garbage collection and transportation.

In order to reduce vehicle movement, a separate unload system—"Ramp"—has been installed around the city. Instead of taking the waste to the process plant, the waste from each ward is brought to the ramp. The waste from smaller capacity vehicles are transferred to the Bulk Refuse Carrier and vehicles with bulk capacity are directly allowed into the processing plant. PMC has transfer stations at Aundh, Kothrud, Yerwada, Hadapsar, Katraj, Ghole Road and Dhole Patil Road.

What about the sanitation programmes?

The corporation has approximately 1500 public toilets. Some of these are being maintained by Sulabh International, some by Shelters and the rest by the Corporation. In 100 locations where open defecation is being practised, we are aiming to provide public toilets.

Source: https://www.cleanindiajournal.com/interview_with_mahesh_pathak-une_creates_a_perfect_model/

Interview 2

"We give awards to companies that pollute less," said Gunter Pauli at a recent conference in Bengaluru organised by Ashoka Trust for Research in Ecology and the Environment. "And we put in jail people who steal less. We forget that polluting less is still pollution." For Pauli, nothing more than zero waste is acceptable: no landfills, no burning, no discharge into the rivers, no e-waste to ship into another country.

Pauli, a Belgian serial entrepreneur who lives in Japan and now works in South Africa—and who has sometimes been called the Steve Jobs of sustainability—has projects in four continents on sustainable farming, on handling urban solid waste, on using the ocean for future food, energy, and so on. In 1992, he built a zero-emissions factory making soaps out of wood. A year later, he found out that it was non-sustainable. This learning resulted in the non-profit Zero Emissions Research and Initiatives (ZERI) in 1994, based in Tokyo. It has since grown around the world. Last year, a report by the University of Pennsylvania on global think tanks ranked ZERI at number seven among those with the most innovative ideas.

On the day Pauli spoke in Bengaluru, a newspaper had reported about a plan to use trains to take garbage out of the city, to a town called Madhugiri about 100 kilometres away for incineration. Bengaluru no longer has space in neighbouring areas to bury garbage. Building incinerators in the city is an idea that will be opposed from the beginning. Residents of Madhugiri are opposing it, too.

Urban solid waste is a seemingly intractable problem in Indian cities, as the garbage mounts and cities run out of space. ZERI has worked on solid waste around the world and has got together some unique ideas for solid waste management. "They have not run out of space," says Pauli.

"They have run out of ideas to make it work. The easiest solution is to ship it out. We first dump it, and then we incinerate it, and we realise that even the incineration is a toxin."

ZERI has worked with the city of Milan in Italy to reduce waste by 90%. The nonprofit's plan is to get the biomass out and do useful things with it. It is working with the city of Milan to recover organic waste. "Milan is now recovering 90 kg of biomass per person per year. It is number one in the world."

Separating organic waste is one thing, making use of it another. Composting does not generate enough money. "In the city, you can't compost," says Pauli, "as composting generates methane. You have to generate value." The trick is to separate the bio-waste further, with an eye on commercial activity. Pauli's prime exhibit is coffee waste, increasing in the country as coffee shops become popular in the cities. "We can take a tonne of coffee waste and produce a tonne of mushrooms."

So separation at the source is the first step, not always easy in a country like India. "Coffee and tea waste are a substrate for mushrooms. It does not make any sense to compost it. You must use it to make food." Similarly, citrus fruit peel can be used to make detergents. "We have mapped hundreds of opportunities. The question is, do we have the entrepreneurs to turn them around?"

Once you generate value for your waste, people look at it differently. Would you throw away your coffee waste if someone pays for it? "This is what we call systemic businesses, the interconnected businesses."

Brazil has eight such factories, according to Pauli. Mexico is starting a factory to process mango seeds and turn it into an additive for bread. "How many mango seeds would India have? Your bread is junk bread. You can make good bread by using mango seeds." Similarly, unused part of vegetables can be used to feed maggots, which in turn can feed the chickens. Maggots can digest almost everything.

Once the organic waste is separated and used up, only a small part is left for composting. Since organic waste is more than half of the city waste, a series of small and networked factories can reduce the total waste significantly. The rest is dominated by two waste streams: plastic and electronic waste. ZERI claims to have a technology to break down plastics using enzymes. The toxic chlorinated plastics are broken down using a combination of enzymes and heat treatment, and then the other plastics are also treated in a similar way. "There are compounds in plastics that industry will not tell you because it is less than 1%," says Pauli. With a judicious use of enzymes, heat and high pressure, plastics can be turned into a fuel that can be burned safely.

The last category is e-waste, an extremely toxic and difficult category to handle. The state-of-the-art method is to evaporate them in a vacuum. "This is a non-starter technology," says Pauli. "We focus on another technology called chelation." It crushes the waste and then allows you to take out each constituent separately for reuse.

Source: http://m.economictimes.com/small-biz/entrepreneurship/the-belgian-entrepreneur-who-has-unique-ideas-for--solid-waste-management/articleshow/57008457.cms

Part D　Supplementary Vocabulary

Solid Waste

agricultural waste 农业废弃物
anaerobic [ˌæneə'rəubɪk]　*adj.* 厌氧的；厌氧菌的；厌氧菌产生的
ash residue 灰渣
circular economy 循环经济
clean production 清洁生产
combustible waste [kəm'bʌstəbl] 可燃废物
corrosivity [kərəu'sɪvɪtɪ]　*n.* 腐蚀性
deplete [dɪ'pli:t]　*vt.* 耗尽；使枯竭；[医] 减液，放血；*vi.* 耗尽；减少
dismantle [dɪs'mæntl]　*vt.* 拆卸；拆开；废除；取消
electronic waste 电子废物
extravagant [ɪk'strævəgənt]　*adj.* 过度的；浪费的；放肆的；大量的
finite ['faɪnaɪt]　*adj.* 有限的；限定的；*n.* 有限之物
flame retardant 阻燃剂
food waste 餐厨垃圾
garbage container 垃圾箱
general waste 一般废物

hazardous waste 危险废物
household garbage 生活垃圾
ignitability [ɪgˌnaɪtə'bɪlɪtɪ]　*n.* 可燃性
illegal dump 非法倾倒
infectiousness [ɪn'fekʃəsnəs]　*n.* 传染性
inorganic waste 无机废弃物
landfill gas 埋填气
landfill site 垃圾填埋区
leachate ['li:tʃeɪt]　*n.* 沥出液，沥出物
environmental awareness 环境意识
noncombustible waste 不可燃废物
occupational hazard 职业病；职业危害，职业冒险
organic waste　有机废物
packaging waste　包装垃圾
pollution prevention　污染防治
radioactive waste [ˌreɪdɪəu'æktɪv] 放射性废物
reactivity　*n.* 反应性
reduce, reuse, recycle 减少，再利用，再循环
residue ['rezɪdju:]　*n.* 残余

solid waste 固体废物
soil flushing ['flʌʃɪŋ] 土壤淋洗
soil washing 土壤冲蚀，土壤冲刷
source reduction 源削减；源减量
special waste 特殊废物
sustainable development 可持续发展
toxicity *n.* 毒性，毒力
toxic waste 有毒废物
transboundary movement 跨界转移
treatment plant wastes 废弃物的处理

trivial ['trɪvɪəl] *adj.* 无价值的；平常的；不重要的；[生] 种的
waste composting field 垃圾堆肥场
waste diversion 垃圾转移
waste layer 垃圾层
waste minimization 废物最小化
waste stream 废物流
zero emission 零排放
zero waste alliance 零废物联盟

Part E Supplementary Knowledge

英文学术论文的题目与摘要

一、英文题名

1. 题名的结构

英文题名以短语为主要形式，尤以名词短语（noun phrase）最常见，即题名基本上由1个或几个名词加上其前置和（或）后置定语构成。例如：Toxicity of vanadium in soil on soybean at different growth stages；Metal distribution in soils of an in-service urban parking lot。短语型题名要确定好中心词，再进行前后修饰。各个词的顺序很重要，词序不当，会导致表达不准。题名一般不应是陈述句，因为题名主要起标示作用，而陈述句容易使题名具有判断式的语义；此外陈述句不够精练和醒目，重点也不易突出。少数情况（评述性、综述性和驳斥性）下可以用疑问句做题名，因为疑问句可有探讨性语气，易引起读者兴趣。例如：Can agricultural mechanization be realized without petroleum?

2. 题名的字数

题名不应过长。国外科技期刊一般对题名字数有所限制。例如，美国医学会规定题名不超过2行，每行不超过42个印刷符号和空格；英国数学会要求题名不超过12个词。这些规定可供我们参考。总的原则是，题名应确切、简练、醒目，在能准确反映论文特定内容的前提下，题名词数越少越好。

3. 中英文题名的一致性

同一篇论文，其英文题名与中文题名内容上应一致，但不等于说词语要一一对应。在许多情况下，个别非实质性的词可以省略或变动。例如：工业湿蒸汽的

直接热量计算，The direct measurement of heat transmitted wet steam。英文题名的直译中译文是"由湿蒸汽所传热量的直接计量"，与中文题名相比较，二者用词虽有差别，但内容上是一致的。

4. 题名中的冠词

在早年，科技论文题名中的冠词用得较多，近些年有简化的趋势，凡可用可不用的冠词均可不用。例如：The effect of groundwater quality on the wheat yield and quality. 其中两处的冠词 the 均可不用。

5. 题名中的大小写

题名字母的大小写有以下 3 种格式。
(1) 全部字母大写。例如：OPTIMAL DISPOSITION OF ROLLER CHAIN DRIVE
(2) 每个词的首字母大写，但 3 个或 4 个字母以下的冠词、连词、介词全部小写。例如：Effects of Lead on Soil Enzyme Activity in Two Red Soils.
(3) 题名第 1 个词的首字母大写，其余字母均小写。例如：Leaching characteristics of vanadium in mine tailings and soils near a vanadium titanomagnetite mining site.
第 2 种和第 3 种格式使用较多。

6. 题名中的缩略词语

已得到整个科技界或本行业科技人员公认的缩略词语，才可用于题名中，否则不要轻易使用。

二、作者与作者单位的英译

1. 作者

中国人名按汉语拼音拼写；其他非英语国家人名按作者自己提供的罗马字母拼法拼写。

2. 单位

单位名称要写全(由小到大)，并附地址和邮政编码，确保联系方便。如果单位机构英译采取缩写，外人不知所云，结果造成混乱。FAO、WHO、MIT 人尽皆知，而 SCU 是 Sichuan University，许多国外学者并不知晓。另外，单位英译一定要采用本单位统一的译法(即本单位标准译法)，切不可另起炉灶。

三、英文摘要

文章摘要是对所写文章主要内容的精炼概括。美国人称摘要为"Abstract"，而英国人则喜欢称其为"Summary"。

通常国际刊物要求所要刊登的文章字数，包括摘要部分不超过 1 万字。而对文章摘要部分的字数要求则更少。因此，写摘要时，应用最为简练的语言来表达论文之精华。论文摘要的重点应放在所研究的成果和结论上。

1. 摘要的要素

(1) 目的——研究、研制、调查等的前提、目的和任务，所涉及的主题范围。

(2) 方法——所用的原理、理论、条件、对象、材料、工艺、结构、手段、装备、程序等。

(3) 结果——实验的、研究的结果、数据，被确定的关系，观察结果，得到的效果、性能等。

(4) 结论——结果的分析、研究、比较、评价、应用，提出的问题等。

2. 摘要的篇幅

摘要的篇幅取决于论文的类型。国际会议要求的论文摘要的字数不等，一般为 200~500 字。而国际刊物要求所刊登的论文摘要的字数通常是 100~200 字。摘要的位置一般放在一篇文章的最前面，内容上涵盖全文，并直接点明全旨。语言上要求尽量简练。摘要通常多采用第三人称撰写。

可采用以下方法使摘要达到最小篇幅：

(1) 摘要中第一句的开头部分，不要与论文标题重复。
(2) 把背景信息删去，或减到最少。
(3) 只限于新的信息。过去的研究应删去或减到最小。
(4) 不应包含作者将来的计划。
(5) 不应包含不属于摘要的说法，如：
"本文所描述的工作，属于……首创"。
"本文所描述的工作，目前尚未见报道"。
"本文所描述的工作，是对于先前最新研究的一个改进"。
(6) 相同的信息不要重复表达。

原文为	应改为
at a temperature of 250 °C to 300 °C	at 250~300 °C
at a high pressure of 1.2 MPa	at 1.2 MPa

(7) 以量的国际单位符号表示物理量单位(例如，以"kg"代替"kilogram")。
(8) 以标准简化方法表示英文通用词(以"NY"代替"New York")。

3. 摘要的时态

英文摘要时态的运用也以简练为佳，常用一般现在时、一般过去时，少用现在完成时、过去完成时、进行时态和其他复合时态基本不用。

大体可概括为以下几点。

(1) 叙述研究过程，多采用一般过去时。

(2) 在采用一般过去时叙述研究过程当中提及在此过程之前发生的事，宜采用过去完成时。

(3) 说明某课题现已取得的成果，宜采用现在完成时。

(4) 摘要开头表示本文所"报告"或"描述"的内容，以及摘要结尾表示作者所"认为"的观点和"建议"的做法时，可采用一般现在时。

一般现在时。用于说明研究目的、叙述研究内容、描述结果、得出结论、提出建议或讨论等。涉及到公认事实、自然规律、永恒真理等，也要用一般现在时。

一般过去时。用于叙述过去某一时刻(时段)的发现、某一研究过程(实验、观察、调查、医疗等过程)。例如：The heat pulse technique was applied to study the stemstaflow of two main deciduous broadleaved tree species in July and August, 1996. 需要指出的是，用一般过去时描述的发现、现象，往往是尚不能确认为自然规律、永恒真理的，而只是当时如何；所描述的研究过程也明显带有过去时间的痕迹。

现在完成时和过去完成时。完成时少用，但不是不用。现在完成时把过去发生的或过去已完成的事情与现在联系起来，而过去完成时可用来表示过去某一时间以前已经完成的事情，或在一个过去事情完成之前就已完成的另一过去行为。例如：Concrete has been studied for many years.

4. 摘要的语态

采用何种语态，既要考虑摘要的特点，又要满足表达的需要。一篇摘要很短，尽量不要随便混用，更不要在一个句子里混用。

主动语态。现在主张摘要中谓语动词尽量采用主动语态的越来越多，因其有助于文字清晰、简洁及表达有力。The author systematically introduces the history and development of the tissue culture of poplar 比 The history and development of the tissue culture of poplar are introduced systematically 语感要强。

被动语态。以前强调多用被动语态，理由是科技论文主要是说明事实经过，至于那件事是谁做的，无须一一证明。事实上，在指示性摘要中，为强调动作承受者，还是采用被动语态为好。即使在报道性摘要中，有些情况下被动者无关紧要，也必须用强调的事物做主语。例如：In this case, a greater accuracy in measuring distance might be obtained.

在多数情况下可采用被动语态。但在某些情况下，特别是表达作者或有关专家的观点时，又常用主动语态，其优点是鲜明有力。

(1) 表示研究目的，常用在摘要之首 In order to…，This paper describes…，The purpose of this study is…

(2) 表示研究的对象与方法 The（curative effect/sensitivity/function）of certain (drug/kit/organ…) was (observed/detected/studied…)

(3) 表示研究的结果：(It proved/The authors found) that…

(4) 表示结论、观点或建议：The authors (suggest/conclude/consider) that…

5. 摘要的人称

原来摘要的首句多用第三人称 This paper…等开头，现在倾向于采用更简洁的被动语态或原形动词开头。例如：To describe…，To study…，To investigate…，To assess…，To determine…，The torrent classification model and the hazard zone mapping model are developed based on the geography information system. 行文时最好不用第一人称，以方便文摘刊物的编辑刊用。

6. 摘要的英文写作风格

要写好英文摘要，就要完全地遵从通行的英文摘要写作规范。其要点如下：

(1) 句子完整、清晰、简洁。

(2) 用简单句。为避免单调，改变句子的长度和句子的结构。

(3) 用过去时态描述作者的工作，因它是过去所做的。但是，用现在时态描述所做的结论。

(4) 避免使用动词的名词形式。如：

正："Thickness of plastic sheet was measured."

误："Measurement of thickness of plastic sheet was made."

(5) 正确地使用冠词，既应避免多加冠词，也应避免蹩脚地省略冠词。如：

正："Pressure is a function of the temperature."

误："The pressure is a function of the temperature."

正："The refinery operates …"

误："Refinery operates…"

(6) 使用长的、连串的形容词、名词、或形容词加名词，来修饰名词。为打破这种状态，可使用介词短语，或用连字符连接名词词组中的名词，形成修饰单元。例如：

应写为："The chlorine-containing propylene-based polymer of high melt index"

而不写为："The chlorine containing high melt index-propylene based polymer"

(7) 使用短的简单的、具体的、熟悉的词。不使用华丽辞藻。

(8) 构成句子时，动词应靠近主语。避免形如以下的句子：

"The decolorization in solutions of the pigment in dioxane, which were exposed to 10 hr of UV irradiation, was no longer irreversible."

改进的句子，应是：

"When the pigment was dissolved in dioxane, decolorization was irreversible, after 10 hr of UV irradiation."

(9) 避免使用那些既不说明问题，又没有任何含意的短语。例如："specially designed or formulated"，"The author discusses"，"The author studied" 应删去。

(10) 不使用俚语、非英语的句子。慎用行话和口语。不使用电报体。

(11) 尽量采用-ing 分词和-ed 分词作定语，少用关系代词 which、who 等引导的定语从句。由于摘要的时态多采用一般过去时，使用关系代词引导的定语从句不但会使句式变得复杂，而且容易造成时态混乱。采用-ing 分词和-ed 分词作定语，在简化语句的同时，还可以减少时态判定的失误。

(12) 尽量使用短句。因为，长句容易造成语义不清；但要避免单调和重复。科技期刊涉及专业多，英文更是不易掌握，各行各业甚至表达方式、遣词造句都有区别。

(13) 应避免一些常见的错误。

冠词。主要是定冠词 the 易被漏用。the 用于表示整个群体、分类、时间、地名以外的独一无二的事物、形容词最高级等较易掌握，用于特指时常被漏用。这里有个原则，即当我们用 the 时，听者或读者已经确知我们所指的是什么。例如：The author designed a new machine. The machine is operated with solar energy. 由于现在缩略语越来越多，要注意区分 a 和 an，如 an X ray.

数词。避免用阿拉伯数字作首词，如：Three hundred Dendrolimus tabulaeformis larvae are collected…中的 Three hundred 不要写成 300.

单复数。一些名词单复数形式不易辨认，从而造成谓语形式出错。

Reference: http://bbs.hcbbs.com/thread-379694-1-1.html
https://wenku.baidu.com/view/5a5887ca71fe910ef12df87c.html?qq-pf-to=pcqq.c2c
https://wenku.baidu.com/view/d95b049f71fe910ef12df854.html?qq-pf-to=pcqq.c2c
https://wenku.baidu.com/view/1ea0dcd3a58da0116c17499d.html?qq-pf-to=pcqq.c2c

Unit 8　Organic Pollution

Part A　Intensive Reading

Pesticides

　　Pesticides can be classified by function and divided into subclasses by structure. However, some of the structural types are used for several functions. For example, carbamates are used as both insecticides and herbicides, as is the organochlorine hexachlorobenzene. Before the development of synthetic pesticides during World War II, other compounds were used, such as arsenicals and nicotine. What they all have in common is that they are intended to control or eliminate undesirable organisms. The emphasis here will be on the insecticides and herbicides because of their wider distribution in the environment.

　　Organochlorine (OC) pesticides are persistent, bioaccumulated, and biomagnified. Thus, their use is avoided now. The organophosphorus (OP) compounds do not have these characteristics but are extremely toxic to mammals. The carbamates are similar to OP compounds but less acutely toxic to mammals. The botanically derived compounds include pyrethrin, which is isolated from the chrysanthemum, and permethrin, a synthetic compound. They are effective and relatively safe. They readily degrade on exposure to sunlight. Half-lives in soil for OCs range from 1 to 12 years; OPs from 0.2 to 0.5 year; carbamates from 0.05 to 1.0 year.

　　The atmosphere is a major route of dispersion. Many pesticides are applied by spraying. The less soluble substances tend to be distributed in the environment adsorbed to particles. As a result, some are transported in significant quantities over long distances attached to dust particles in the atmosphere or as suspended solids in aquatic systems.

　　Many pesticides are degraded by photooxidation and by microbes in the environment into other toxic substances. DDT is converted to DDE, aldrin to dieldrin, and heptachlor to heptachlor epoxide.

Toxic Effects

　　Most insecticides act by interfering with the nervous system. For example, DDT binds to lipoproteins in axon membrane, holding a gate open for sodium.

Organochlorines in general stimulate the nervous system, causing irritability, tremor, and convulsions.

Exposure to OPs was observed in both human and animal tests to result in myopathy (damage to muscle tissue), including necrosis of skeletal muscles. This is thought to be caused by the high levels of acetylcholine resulting from inhibition of acetylcholinesterase. Symptoms include muscle tenderness. Adverse psychological disturbances have been identified which last from six months to a year. These include depression, nightmares, and emotional instability.

There is positive interaction between many OPs. One of the strongest is a 100-fold increase in the combined toxicity of malathion and tri-o-cresyl phosphate (TOCP). Some pesticide formulations include potentiators to increase their effects, although this increases mammalian toxicity. Antagonistic effects occur as well. Substituted urea herbicides reduce the toxicity of parathion, probably by inducing MFO detoxification.

OP and carbamates are not carcinogens, in general, except some that are chlorinated, or some like carbaryl, which can be nitrosated to a carcinogenic form. Carbaryl produced effects attributed to decreased renal resorption in human experiments. Paraquat causes edema, hemorrhage, and fibrosis in the lungs when inhaled. Pyrethrum is associated with hypersensitivity, contact dermatitis, and asthma. OCs are hepatotoxic and induce microsomal monooxygenases. Some pesticides from all the major groups have an effect on the immune system. These include malathion, methylparathion, carbaryl, DDT, paraquat, and diquat.

Herbicides act on plants by affecting either growth or photosynthesis. The phenoxyacids simulate natural auxins (plant growth hormones). Urea herbicides inhibit photosystem II, preventing ATP and NADPH formation. Diquat and paraquat inhibit photosynthesis. Some are selective for either monocots (grassy plants) or dicots (broadleaf).

The manufacture of the pesticide 2,4,5-T produces 2,3,7,8-tetrachlorodibenzo-p-dioxin (TCDD) as a trace impurity. This is one of the most toxic substances known. The manufacture and use of 2,4,5-T and related herbicides has resulted in some of the most notorious environmental contamination incidents known, which involve TCDD.

Dinitrophenols are used for weed control. They uncouple oxidative phosphorylation in respiration, causing metabolic activity to increase out of control. Effects of acute exposure in humans include the sensation of heat, and rapid breathing and heart rate. Chronic symptoms include anxiety, sweating, thirst, and fatigue. Paraquat causes lung, kidney, and liver damage and can cause pulmonary fibrosis even from routes other than inhalation.

Ecosystem Effects

Biomagnification is a problem mostly with organochlorine pesticides. The classic case is DDT and its metabolite DDD. By their very nature, pesticides can produce pronounced effects on populations and ecosystems. These include many that were dramatically publicized by Rachel Carson in her book Silent Spring, including fish and bird kills, development of insect resistance, and actually causing an increase in some pests by destroying their predators. In the latter case, the target species is sometimes the beneficiary of a pesticide application, if its predators and competitors are more strongly affected. An example of this counterproductive situation is the application of pesticide to control boll weevils and leafworms infesting cotton farms in Central America. From 1950 to 1955 the number of pesticide applications had to increase from few or none to 8 to 10 per season. By 1960, 28 applications per season were needed. Over the same period, the number of pest species increased from the original two in 1950, to five in 1955, and then eight in 1960. Many similar examples exist.

Many pesticides, such as insecticides, fungicides, and fumigants, target soil organisms and have their greatest environmental effects on nontarget organisms in the soil. Susceptible invertebrates include insects, earthworms, slugs, and gastropods. Some of these organisms contribute to topsoil formation. Insecticides applied to natural systems, such as for the control of forest pests, can reduce the species diversity within the invertebrates.

Among the terrestrial vertebrates, avian species are especially susceptible to the insecticides. LD_{50} values are often less than 100 mg/kg. However, the sublethal effects are of greatest concern. In particular, bird reproduction is very sensitive to organochlorine insecticides. The best-known effect is the thinning of eggshells due to DDT, its metabolite DDE, and dieldrin, among others. These prevent the embryos from surviving to hatch. However, many other reproductive effects are also found, including direct embryo toxicity and aberrant parental behavior such as destruction of eggs. The greatest concern with these effects is with birds of prey, because the organochlorine pesticides are biomagnified. Species that have experienced serious population declines include the bald eagle (Haliacetus leucophalus) and the osprey (Pondion haliatetus).

DDT has not been found to seriously affect wild mammals. However, other organochlorine pesticides, especially the cyclodienes such as heptachlor, dieldrin, and endrin, have resulted in mortalities. Although the cholinesterase inhibitors are very toxic to mammals, no significant effects have been observed in the wild. Development of resistance has been observed in a variety of aquatic invertebrates.

Aquatic ecosystems have been affected both by direct application of pesticides, such as for control of mosquitoes, and by runoff from agricultural, forest spraying, and so on. Arthropods such as crustaceans are very susceptible to insecticides. Heavy killings of crabs and shrimp have occurred in marshes sprayed with organochlorines. The microcrustaceans are an important component of the food chain and are also affected heavily. Mollusks and annelids, on the other hand, are relatively tolerant. Fish tend to be very susceptible to the organochlorines, especially the cyclodienes. Some warm-water, rapidly reproducing species of fish in heavily sprayed areas of Mississippi have become resistant to levels of some of the organochlorines 100 times higher than that which is normally toxic.

Source: Vaccari D A, Strom P F, Alleman J E. Environmental Biology for Engineers and Scientists[M]. Hoboken :John Wiley & Sons, Inc., 2006.

Words and phrases

pesticide ['pestɪsaɪd]　　*n*. 杀虫剂，农药
carbamate ['kɑːbəˌmeɪt]　　*n*. [医]氨基甲酸酯(类)
insecticide [ɪn'sektɪsaɪd]　　*n*. 杀虫剂
herbicide ['hɜːbɪsaɪd]　　*n*. 灭草剂
arsenical [ɑː'senɪkəl]　　*n*. 砷剂；*adj*. 砷的，含砒素的
acutely [ə'kjuːtli]　　*adv*. 尖锐地；剧烈地
botanically [bə'tænɪklɪ]　　*adv*. 植物学地
pyrethrin [paɪ'riːθrɪn]　　*n*. 除虫菊素
chrysanthemum [krɪ'sænθəməm]　　*n*. 菊花；菊属
permethrin [pə'meθrɪn]　　*n*. 合成除虫菊酯
spray ['spreɪŋ]　　*n*. 喷雾；*v*. 喷；向…扫射；往…上撒
photooxidation [fəʊtɒksɪ'deɪʃən]　　*n*. 光氧化
axon ['æksɒn]　　*n*. 轴突
membrane ['membreɪn]　　*n*. 膜

necrosis [ne'krəʊsɪs]　　*n*. 坏疽，骨疽
acetylcholine [ˌæsɪtɪl'kɒliːn]　　*n*. 乙酰胆碱
acetylcholinesterase [æsɪtɪlkəʊlɪ'nestəreɪs]　　*n*. 乙酰胆碱酯酶
malathion [ˌmæləˈθaɪɒn]　　*n*. 马拉硫磷
potentiator [pə'tenʃɪeɪtə]　　*n*. 增效剂
antagonistic [ænˌtægə'nɪstɪk]　　*adj*. 敌对的；对抗性的；反对的
urea [jʊ'riːə]　　*n*. 尿素
parathion [ˌpærəˈθaɪɒn]　　*n*. 对硫磷
carcinogen [kɑː'sɪnədʒən]　　*n*. 致癌物
carbaryl ['kɑːbərɪl]　　*n*. 甲萘威
nitrosated [naɪt'rəʊzət]　　*adj*. 硝基化的
paraquat ['pærəkwɒt]　　*n*. 百草枯
edema [ɪ'diːmə]　　*n*. 水肿
hemorrhage ['hemərɪdʒ]　　*n*.(尤指大量的) 出血，失血；*vi*. 大出血
fibrosis [faɪ'brəʊsɪs]　　*n*. 纤维化，

纤维症
hypersensitivity [ˌhaɪpəˌsensə'tɪvəti] *n.* 超敏反应
hepatotoxic [ˌhepətəʊ'tɒksɪk] *n.* 肝毒性
monooxygenase [mɒnə'ɒksɪdʒɪneɪs] *n.* 单加氧酶
auxin ['ɔːksɪn] *n.* 生长素
monocot ['mɒnəkɒt] *n.* 单子叶植物
dinitrophenol [daɪnaɪtrəʊ'fiːnɒl] *n.* 硝基苯酚
predator ['predətə(r)] *n.* 食肉动物，捕食者
counterproductive [ˌkaʊntəprə'dʌktɪv] *adj.* 反生产的，使达不到预期目标的
infest [ɪn'fest] *vt.* 骚扰；大批出没，成群出现；在…上寄生
gastropod ['gæstrəpɒd] *n.* 腹足类
embryo ['embriəʊ] *n.* 胚胎
osprey ['ɒspreɪ] *n.* 鱼鹰
cyclodiene [saɪkləʊ'daɪiːn] *n.* 烷基酚
mollusk ['mɒləsk] *n.* 软体动物
annelid ['ænəlɪd] *n.* 环节动物
involuntary urination 小便失禁
adverse psychological disturbance 不良的心理扰动
thyroid tumor 甲状腺肿瘤
oxidative phosphorylation 氧化磷酸化
renal resorption 肾重吸收
trace impurity 痕量杂质
terrestrial vertebrate 陆生脊椎动物

Questions

1. How do most insecticides work?
2. How do most herbicides work?
3. Who is most sensitive to insecticides in terrestrial vertebrates?
4. Give some examples of abiotic factors that influence the effects of pesticides on aquatic organisms.
5. Does DDT have a great impact on wild mammals?

Sentence-making

1. pesticide, insecticide, pollution
2. spraying, runoff, sediment
3. carcinogen, epidemiological, environment
4. metabolic, accumulation, bioremediation
5. acute, chromic, toxicity

Part B Extensive Reading

How to Clean a Beach

As oil-spill specialists continue to tackle the Prestige slick, they are drawing on knowledge from decades of clean-up operations. John Whitfield reports from Spain's Galician coast.

From the coast road, the beaches of Lira seem as they should be: yellow sand and blue sea. But walk down to the tide's edge and things change. A whiff of petrol taints the sea spray. Water in rock pools has an oily sheen and boulders that should be wet and slippery have a tacky, tarry coating. After the oil-tanker Prestige spilt her cargo last November, these coves in Galicia, in northwest Spain, were a metre deep in a mixture of oil and sea water known to pollution specialists as "chocolate mousse". "There was no ocean, only oil," says Pablo Garcia, manager of the Stolt Sea Farm, an aquaculture company in Lira, the area that became known as Ground Zero of the spill.

The Prestige is the latest exhibit in the tanker hall of infamy. But while each new incident brings environmental destruction and financial loss, it also improves our understanding of how to deal with oil spills. This knowledge is hard-won — aggressive clean-ups have sometimes caused more damage than the oil. Government priorities can also clash with those of scientists. But such difficulties apart, a rough consensus on how to juggle the political, economic and ecological issues involved in clearing up oil spills has begun to emerge.

Oil is much less damaging at sea than on shore, so the best option is to suck and skim a slick off the water using specially equipped ships, or break it up with chemical dispersants. Booms can also be used to protect the coastline. But the sea around the Prestige was too rough and much of the coast too exposed for booms to work, so there was little that could be done except watch the oil wash up.

The Human Touch

When oil arrives onshore, the question becomes how best to save affected plants and animals while minimizing damage to the surrounding ecosystem, and without running up a huge bill. In large spills, leaving nature to do the job is a bad idea. Even the oiliest shore will return to normal, but without human intervention this can take a long time. In 1974, the Metula spilled 50,000 tonnes of oil into the Strait of Magellan at the southern tip of South America. Because of the remoteness of the region and the rough seas, no clean-up was mounted, and patches of asphalt-like residue stain the rocks to this day. "It looks like a cheap driveway," says David Page, a chemist at Bowdoin College in Brunswick, Maine.

In the case of the Prestige, the volume of oil spilt and the wildlife, fishing and tourist value of the Galician coast demanded action. The first priority in a clean-up is clear — remove oil from the beaches as quickly as possible. If washed back out to sea, or buried in the beach, oil can do more damage some other time or place.

Cleaning beaches is ideally done manually. People with shovels are the only tools sensitive enough to remove the oil while protecting the ground beneath. Only the human eyes can distinguish patches of oil from the clean areas in between: on some of the Galician beaches, oil-coated rocks and apparently unaffected rocks sit side by side. And people can work on isolated rocky shores where heavy machinery cannot go.

But at Lira in early December, manual labour was having no effect. "We got the full load of a couple of tanks of the Prestige — 15,000 tonnes on two kilometres of coast," says Garcia. "You'd see guys working manually, and at the end of the day, the area occupied by the slick was the same."

To decide what to do next, specialists combined their experience with local knowledge in an assessment process known as net environmental benefit analysis. Factors taken into account include environmental considerations, such as whether an oiled beach is home to a breeding colony of seals or seabirds. Socioeconomic considerations also come into play: local people may rely on nearby shellfish beds, for example, or an affected beach could be a tourist destination. And practical realities, such as how much a particular clean-up option will cost, and how easy it will be to implement, are also assessed.

Every intensive clean-up option has its drawbacks. When high-pressure hot water was used to scrub the Alaskan shoreline oiled by the Exxon Valdez spill in 1989, beaches that got this treatment recovered more slowly than those that did not, although conditions seemed about equal after three years. This technique is now used less often, says Alan Mearns, a marine ecologist with the Hazardous Materials Response Division of the US National Oceanic and Atmospheric Administration (NOAA) in Seattle, Washington. "We have a go-easy policy on using rigorous methods," he says. "The coastguard understands that you don't have to go in with all guns blazing." In Spain, high-pressure water has so far been used only on man-made structures such as jetties and harbour walls.

Chemical cleaners also do damage. When the Torrey Canyon ran aground off the south coast of Cornwall, UK, in March 1967, oiled beaches were sprayed with 10,000 tonnes of powerful solvents and detergents, including industrial degreasers. The chemicals were more toxic than the oil: many seashore invertebrates died, and the nutrients in the dispersants caused an explosion of seaweed growth.

Dispersants are much milder now, but they can still be used inappropriately. Peter Dyrynda, a marine biologist at the University of Wales Swansea, UK, says that after the Sea Empress ran aground off the southwest coast of Wales in 1996, dispersants were used on some patches of oil but not washed off. The resulting mix was more toxic than either oil or dispersant alone and killed animals that had survived until then. Such problems, as well as the fact that dispersants are ineffective against thick oil, have prevented their use in Galicia.

Gentle Clean

A gentler option is bioremediation, which involves using fertilizer to speed the growth of naturally occurring oil-digesting bacteria. But this won't work on every spill, says Richard Swannell, a bioremediation specialist with Momenta, a consultancy company near Oxford, UK, that works with the British government. The oily shore must be sheltered, otherwise the sea will wash the oil and bacteria away. Oil buried in sediments can't be digested. And if the natural bacterial growth is limited by temperature rather than by nutrients, the treatment will have no effect.

The oil must also be biodegradable: light crude oils of the type spilt by the Exxon Valdez are broken down easily, but heavier types contain compounds that microbes find indigestible. Finally, sites must be secluded, so that they do not offend the senses of local people and tourists during the several months the bacteria need to work. The remote Alaskan shores polluted by the Exxon Valdez fit the bill, but many others do not. "For marine spills, bioremediation is a niche market," Swannell concludes. Spanish researchers began bioremediation experiments on oiled beaches late last month, although the Prestige's thick, poorly biodegradable oil might not respond well to the treatment.

In Galicia, local people wanted heavy machinery to come in and remove large quantities of oil quickly. But this has its own environmental costs. There were no tracks to the worst-affected beaches and building them would have harmed the surrounding landscape. "If you've got a site that no one visits, then it's ideal to leave for natural clean-up. If it's an area that people go to all the time you can't do that, because people will be getting oil on them," says Rob Self of Oil Spill Response, a company in Southampton that worked out of the command centre in La Coruña to advise the Spanish authorities on the clean-up operation. But at Lira, oil on the surface of the beaches was likely to be washed into the pipe that supplied water to Garcia's fish farm. In the end, the desire to protect the farm tipped the balance in favour of building tracks for bulldozers and earthmovers, which scooped up the chocolate mousse from the beach.

Heavy machinery has not been used in all areas affected by the Prestige spill. Salt-marshes and estuaries are such delicate terrain that almost any activity does more damage than the oil. After the Amoco Cadiz disaster in 1978, when nearly a quarter of a million tonnes of oil were spilt off the coast of Brittany, in northwest France, heavy equipment was sent into some polluted salt-marshes, where it scraped up the top half-metre of sediment. Twelve years later, these areas had still not recovered, whereas the oiled marshes that went uncleaned seemed in good shape.

In Galicia, some marshy areas were placed completely off-limits, even to people, when it became clear that volunteers were cleaning with excessive gusto. "People were pushing the oil into the substrate, and that has more of an effect than if you just left it," says Self. "In the end we had to close the site."

Using a combination of manual labour and heavy machinery, the Prestige spill has now moved from what Self calls the emergency phase — high profile, high pressure — to a long-term painstaking project. Cleaning beaches, for example, is a sisyphean task. At Carnota, one of the most heavily affected areas of Galicia, the high-water mark on the beach in February continues to be marked by a chain of thumbnail-sized oily gobbets. Dozens of volunteers and soldiers work their way along six kilometres of sand on hands and knees, picking up the small lumps with what look like wallpaper scrapers. At other beaches nearby, people sift the sand and comb seaside plants by hand to remove the oil and stop it becoming buried.

Perhaps the most difficult decision, and one usually taken by local politicians, is when to stop. As the coastline becomes cleaner, the clean-up starts to cost more for progressively less reward and more environmental damage. On rocky shores, once the bulk of the oil is recovered, the decision often comes down to whether the beach is an eyesore. Once workers have done all they can with shovels and pompoms — balls of plastic strips that soak up oil — stubborn patches still remain under rocks and in crevices. Moving oily boulders down the beach into the surf can accelerate the natural cleaning process, as can flushing — pumping sea water over the shore and sifting the oil out of the run-off.

As the Prestige clean-up continues into spring and summer, researchers are starting to try to assess the spill's effects. Marine toxicologist Ricardo Beiras of the University of Vigo in Galicia hopes to produce data on damage to local fisheries on which compensation claims can be based. Researchers will also be called on to pronounce when the coast has recovered. This doesn't necessarily mean that no trace of oil remains — old, weathered oil is not very toxic, and oil locked in sediments may not harm organisms.

But the definition of recovery is disputed. The owners of the Exxon Valdez are still fighting with Alaskans over whether the area hit by the spill is still suffering. Page, whose funding comes partly from Exxon, thinks it isn't. "With all spills there are places you can go back to and dig and find a deposit," he says. "The question is whether those isolated remnants are biologically relevant. And the answer is no, they're not."

Others disagree. Stanley Rice of NOAA's Alaska Fisheries Science Center in Auke Bay says the oil is still damaging animals that live and feed on the seashore, such as otters and salmon. "These damages are new and continuing from the remaining oil, and not just a slow recovery from the original hit," says Rice. To try to minimize such chronic effects, he advocates a swing towards more intensive cleaning. "I would push for a more aggressive clean-up, realizing that for the short term you are going to suffer more damage."

Political Science

Resolving issues such as these will require more research, but studies of spills are often only carried out in the fraught atmosphere after a disaster. As a result, clean-up efforts are not as well informed or coordinated as they could be. "We don't clean up spills as well as we should be able to," says Ian White, managing director of the London-based International Tanker Owners Pollution Federation, which responds to spills around the world. "Spills take on a political significance that's hard to control."

Research tends to be improvised and opportunistic, as local scientists drop what they were doing and start studying the pollution. "In the early days of a major spill things are terribly chaotic," says Dyrynda. "Scientific study is not a top priority." Like many others, Dyrynda had oil-pollution research thrust upon him when the Sea Empress ran aground close to where he works. Scientists in northwest Spain are having the same experience, and once again the research is not running smoothly. "There are different individual studies, but they're not coordinated. People are diverting resources from other projects," says Beiras.

But despite the gaps in our knowledge, scientists say that the major threat to the coast, should another spill occur, is not lack of research but of government preparedness. There have been six major oil spills in Galicia. The region is also a hotbed of marine science, with four institutes in Vigo alone. But such experience doesn't guarantee that government officials will talk to local experts. "There's not a lack of knowledge, there's a lack of communication," says plankton ecologist Pablo Serret, of the University of Vigo.

Many Spanish researchers accuse their government of ignoring scientific advice in their handling of the spill — particularly in its decision to tow the Prestige out to sea rather than into port — and of seeking to play down the incident's severity. Research into clean-up methods may be coming together. But as the strength of feeling among Spanish researchers attests, good science is of limited use unless scientists have the backing of politicians. Without that, the damage from oil spills risks going unchecked. "We haven't learnt from the past," says Beiras.

Source: Whitfield J. Oil spills: how to clean a beach[J]. Nature, 2003, 422(6931):464-466.

Words and phrases

cargo [ˈkɑːɡəʊ]　　n.(船或飞机装载的)货物；负荷，荷重
[例句] A fishing vessel and a cargo ship collided in rough seas.
aquaculture [ˈækwəkʌltʃə(r)]　　n. [水产] 水产养殖；水产业
[例句] Another is a shortage of land and freshwater for use in aquaculture.
infamy [ˈɪnfəmɪ]　　n. 声名狼藉；恶行
[例句] The important question it raises is whether accidental infamy is a good thing.
dispersant [dɪsˈpɜːsənt]　　n. 分散剂；悬浮剂；扩散剂；料浆稀释剂
[例句] The aqueous polyaminoester emulsion was gained with water as the dispersant.
onshore [ˈɒnʃɔː(r)]　　adj. 陆上的；吹向陆地的；向岸的；adv. 向着海岸；在陆上
[例句] Greater integration of the onshore and offshore foreign exchange market would make these pressures more immediate.
asphalt [ˈæsfælt]　　n. 沥青；柏油；v. 用柏油铺(路)
[例句] Ultraviolet radiation is one of the main factors inducing asphalt binder aged in the field in-service.
bioremediation [baɪərɪmiːdiˈeɪʃn]　　n. 生物修复；生物复育；生物降解
[例句] The invention discloses a bioremediation method for petroleum contamination and its special strains.
microbe [ˈmaɪkrəʊb]　　n. 细菌，[微] 微生物
[例句] The oil is attacked by naturally occurring microbes which break it down.
substrate [ˈsʌbstreɪt]　　n. 基质；基片；底物层(等于 substratum)；酶作用物

[例句] The pink coralline algae is an early sign of recovery: The algae form a substrate on which new coral can attach itself and grow.
oily [ˈɔɪli]   *adj.* 含油的；油污的；涂油的；(质地、味道、气味、形态)像油的
[例句] The UK Food Standards Agency advises consumers eat at least one portion of oily fish a week.
sediment [ˈsedɪmənt]   *n.* 沉积；沉淀物
[例句] Ocean plants buried in sediment can help reveal Earth's temperature thousands of years ago.
ecologist [iˈkɒlədʒɪst]   *n.* 生态学家；生态保护论者
[例句] The reinstatement work should include the supervision or monitoring of a competent ecologist.

oil-spill 水面浮油；溢油
oil tanker 油轮
environmental destruction 环境破坏
slick off 浮油
human intervention 人为干涉
residue stain 残留污渍
assessment process 评价程序，评估过程
breeding colony 繁殖集落；繁殖集群；集群繁殖
salt marsh 盐碱滩

Questions

1. What's the best option of cleaning up the ocean with oil spill?
2. What's the process of "clean-up" in the case of the Prestige?
3. What're the drawbacks of each intensive clean-up option?
4. Which oil spills are fitted in bioremediation?
5. How can scientists participate in cleaning a beach?

Part C   Dialogue

Dialogue 1

Aaron: Dr. Baldwin, are the Great Lakes contaminated by some toxic compounds?
Baldwin: Yes, the rivers that serve as the lifeblood of the Great Lakes abound with organic waste compounds, including pollutants from car exhaust, factory smokestacks and tar, insecticides and herbicides and detergent ingredients.
Aaron: So where are the most serious area in the Great Lakes?
Baldwin: Three Michigan rivers are among the worst of the worst for the compounds: the Clinton and Rouge rivers in Southeast Michigan and the St. Joseph in Southwest

Michigan, according to a recently published study by the U.S. Geological Survey — the most comprehensive look yet at organic waste compounds in Great Lakes tributaries. The Clinton, Rouge and St. Joseph rivers top the list with the highest number of organic waste compounds found in water sampling USGS conducted throughout the Great Lakes basin in five states over four years, from 2010 to 2013.

Aaron: God! It sounds very serious. But are they harmful for the ecosystem in rivers?

Baldwin: The compounds are known to disrupt the endocrine systems — which produce hormones — of fish and other organisms. They can act like the hormone estrogen and lead to reproductive disruption, including causing male fish to turn into females. Some of the samples we collected in the rivers had concentrations of these estrogenic chemicals comparable to concentrations in effluent from wastewater treatment plants.

Aaron: So they will be accumulated by our human beings from food chain.

Baldwin: Yeah, you are right. Even low levels of the compounds, affecting the smallest shrimp-like invertebrate creatures in the water, cause a ripple effect up the aquatic food chain and could ultimately impact Michigan's prized, multibillion-dollar fishing tourism industry.

Aaron: Where are the compounds come from?

Baldwin: The most prevalent compounds found in the Clinton, Rouge and St. Joseph rivers at relatively high concentrations were so-called polycyclic aromatic hydrocarbons, or PAHs. According to the federal Agency for Toxic Substances and Diseases Registry, they're a group of chemicals formed during incomplete burning of coal, oil, gas, wood, garbage or other organic substances. They're often found in mixtures with gasoline and diesel exhaust or emanating in soot from industrial smokestacks.

Aaron: How can we do to decontaminate these pollution in the Great Lakes?

Baldwin: The findings of the USGS study can help explain to residents and communities the importance of green infrastructure and other management practices to mitigate the impacts. We work with our communities to provide education and presentations on storm-water issues that deal with these compounds and how residents can help lower these numbers.

Aaron: Thanks for your patient expound. Wish you have a good day!

Baldwin: Oh, that's my pleasure.

Reference: http://www.freep.com/story/news/local/michigan/2016/07/30/study-michigan-rivers-organic-pollutants/87317230/

Dialogue 2

Leo: Hey, Jemma, can you help me to review my Environmental Chemistry exam?

Jemma: Sure, which part is your weakness?

Leo: Oh god! It is part 4, the Organic Pollution.

Jemma: Ok, let's start with this part. The first question, can you briefly describe the organic pollution?

Leo: Emm... I see. Organic pollution is that contaminants in an environment that can be oxidized (biodegraded) by microorganisms. It occurs when large quantities of organic compounds, which act as substrates for microorganisms, are released into water substance.

Jemma: Oh, you have a great answer, it didn't bother you. I'm going to give you a difficult question. So can you tell me why these organic compounds are toxic?

Leo: Ok, let me think a moment. During the decomposition process the dissolved oxygen in the receiving water may be used up at a greater rate than it can be replenished, causing oxygen depletion and having severe consequences for the stream biota. Organic effluents also frequently contain large quantities of suspended solids which reduce the light available to photosynthetic organisms and, on settling out, alter the characteristics of the river bed, rendering it an unsuitable habitat for many invertebrates. And, the toxic ammonia is often present.

Jemma: What are the major categories of organic pollutants?

Leo: Organic pollutants consist of proteins, carbohydrates, fats and nucleic acids in a multiplicity of combinations.

Jemma: What are the origins of organic pollutants?

Leo: Organic pollutants originate from domestic sewage (raw or treated), urban run-off, industrial (trade) effluents and farm wastes. Sewage effluent is the greatest source of organic materials discharged to freshwaters.

Jemma: Please give an example to describe this discharge mode.

Leo: In England and Wales there are almost 9000 discharges releasing treated sewage effluent to rivers and canals and several hundred more discharges of crude sewage, the great majority of them to the lower, tidal reaches of rivers or, via long outfalls, to the open sea. It has been assumed, certainly incorrectly, that the sea has an almost unlimited capacity for purifying biodegradable matter.

Jemma: The organic pollution is a big problem to our environment. Please briefly list the effects of organic effluents on receiving waters.

Leo: When an organic polluting load is discharged into a river it is gradually eliminated by the activities of microorganisms in a way very similar to the processes in the sewage treatment works. This self-purification requires sufficient concentrations of oxygen, and involves the breakdown of complex organic molecules into simple in organic molecules. Dilution, sedimentation and sunlight also play a part in the process. Attached microorganisms in streams play a greater role than suspended organisms in self-purification. Their importance increases as the quality of the effluent increases since attached microorganisms are already present in the stream, whereas suspended ones are mainly supplied with the discharge.

Jemma: The last question is what are the effects on the biota?

Leo: Organic pollution affects the organisms living in a stream by lowering the available oxygen in the water. This causes reduced fitness, or, when severe, asphyxiation. The increased turbidity of the water reduces the light available to photosynthetic organisms. Organic wastes also settle out on the bottom of the stream, altering the characteristics of the substratum.

Jemma: Ok, that's all. You have wonderful answers about these questions in the part 4. I think it is a piece of cake for you. So, wish you get a good grade.

Leo: Oh, you are so nice, thank you, Jemma. Wish you have a good day.

Reference: http://www.lenntech.com/aquatic/organic-pollution.htm

Part D Supplementary Vocabulary

Organic Pollution

acridine orange direct count（AODC） 吖啶橙直接计数法
agricultural pollution source 农业污染源
alicyclic [ˌælɪˈsaɪklɪk] *adj.* 脂环族的；*n.* 脂环族
aliphatic [ˌæləˈfætɪk] *adj.* 脂肪族的，脂肪质的
benzopyrene [ˈbenzəpaɪriːn] *n.* 苯并芘
cluster analysis 群分析；聚类分析
endocrine disrupting compounds **(EDCs)** [ˈendəʊkrɪn] 内分泌干扰物
environmental background value 环境背景值
environment capacity 环境容量
environmental quality 环境质量
environmental quality standard 环境质量标准
eutrophication 富营养化
fertilization [ˌfɜːtəlaɪˈzeɪʃn] *n.* 施肥
flocculation [ˌflɒkjʊˈleɪʃən] *n.* 絮凝，絮结产物

harmful algal bloom（HAB） 水华
hydrophobic [ˌhaɪdrə'fəʊbɪk] *adj.* 疏水性的
hydrocarbon [ˌhaɪdrə'kɑːbən] *n.* 碳氢化合物，烃
membrane technique 膜技术
mixed liquor suspended solids 混合液悬浮固体
nitrogen oxide 氮氧化物
non-point source pollution 非点源污染
oil extraction 油萃取；原油开采；石油抽提
organic pollution 有机污染物
organic loading rate（OLR） 有机负荷
olfactory index 嗅觉指标
organic solvent *n.* 有机溶剂
organohalogen [ɔːgænəʊ'hælədʒən] *adj.* 有机卤素的
pharmaceuticals and personal care products（PPCPs） 药物和个人护理用品
pharmaceutical waste 医药废物
persistent organic pollutant 持久性有机污染物
petroleum refining 石油加工
phenanthrene [fə'nænθriːn] *n.* 菲（用于合成染料和药物）
pressurized liquid extraction 加压湿法萃取
primary treatment 一级处理
point pollution source 点污染源
pollution control 污染控制
pollution coefficient 污染系数
pollution indicating organism 污染指示生物
pollution intensity 污染强度
pollution level 污染水平
pollution load 污染负荷
pollution monitoring 污染监测
pollution source 污染源
polycyclic aromatic 多环芳烃
polychlorinated biphenyl 多氯联苯
polymer waste 聚合物废物
polysaprobic zone 多污带
secondary treatment 二级处理
soil vapor extraction ['veɪpə(r)] 土壤气相抽提
synthetic deterge surfactant 表面活性剂(的)
synthetic detergent [dɪ'tɜːdʒnet] 合成洗涤剂
tar production 焦油生产
synthetic textile fiber ['tekstaɪl] 合成纤维纺织品
suspended solid 悬浮物
tertiary treatment 三级处理
total organic carbon 总有机碳
total organic matter（TOM） 总有机物
turbidity 浊度
ultrasound ['ʌltrəsaʊnd] *n.* 超声；超声波
volatile organic compounds（VOCs） 挥发性有机化合物
wet air oxidation 湿式氧化法；湿式空气氧化；湿式氧化

Part E Supplementary Knowledge

如何写英文学术论文的引言

1. 概述

    引言指南 1：吸引读者和提供背景信息

  引言的目的有两个：吸引读者阅读论文和提供足够的背景信息以使读者了解早期发表的相关研究论著。引言通常也对全文内容进行简要概述。引言汇总可少量地重复摘要中的内容。

2. 内容和组织

    引言指南 2：遵循"漏斗形"结构
     研究背景
     未知/存在问题
     提出问题/研究目的/实验方法
     选择性内容：结果/结论/意义
    引言指南 3：保持引言简短

  读者对论文中哪些位置出现何种特定的内容有相对固定的期待，并基于这些内容所处的位置来理解文本。如果作者能意识到位置的作用，就可以根据各种信息的重要程度进行突出和强调，从而更好地引导读者阅读和理解论文。通常情况下，读者期望引言部分有一个标准的结构："漏斗形"，以广泛的背景信息开始，然后收缩至本文讨论的问题。

  引言应该尽可能地简短，但要包含所有必需的信息以引导本文的阅读。理想情况下，一篇期刊论文的引言在双倍行距时为 1~2 页(250~600 个词)。查核拟投稿期刊的作者须知以确认引言的篇幅是否满足期刊的要求。

  大多数基础科学的研究论文都是调查性的，即：这些论文多基于作者试图回答的特定研究问题或试图检验的假设。这类论文的引言应包含以下几部分：

  (1) 背景　　　　　　广泛的和专业的背景信息及本领域已有的研究
  (2) 未知/问题　　　　前人研究的问题及该领域的未知事实
  (3) 提问/研究目的　　本研究的贡献
  (4) 实验方法　　　　本研究采用的方法

  引言应该是漏斗形，从广阔的一般性背景到研究主题的专业内容和相关问题，直至论文的研究问题和实验方法。尽管不是必须，建议作者在引言中指出结果和结论以及论文的总体重要性，以此作为引言的结束。如果引言中包括主要结果和结论，就应将其置于引言的最后。引言中包含结果和结论有助于满足读者的预期并且可以引导读者理解论文。然而，如果论文讨论的是领域内某个有争议的论题，可考虑在引言中不涉及主要结果和结论，以鼓励尽可能多的读者继续阅读全文和论证。

3. 引言的组成

1) 背景

引言指南 4：向读者提供相关的背景信息，但不要综述文献

引言以提供背景信息为开头。有关背景信息的篇幅取决于潜在读者对论文主题的熟悉程度。可以基于较广泛的范围提供研究工作的一般性背景，然后阐述研究主题的各方面，包括已有的研究及相关认识的讨论。

需要注意的是，撰写研究论文时不要综述研究主题。有关研究内容的简短总结已足够。

2) 未知/问题

引言指南 5：指出未知或问题

当一般性研究背景和相关方面阐述完毕之后，应当给出本研究中存在的问题或未知方面。应直接、清楚地表述未知方面，如"X is unknown"或"Y is unclear"。也可直接使用特定的短语表达未知方面，如"has not been established"或"has not been determined"。也可选择提出建议或某种可能性的方式来表达未知方面（"Previous findings suggest that…"）。

应使用客观的语气评论前人的工作，避免使用具有对立含义的短语。

不适合	更好
does not seem to understand	The results of study X have been questioned.
failed to…	One study found A, another study found B.
made the mistake of …	Findings on X are controversial.
used improper methods…	Although A showed X, our results do not agree…

3) 问题/目的

引言指南 6：准确地指出中心点(问题/目的)

研究论文中最重要的因素是研究问题或研究目的。问题/目的是引言或论文的"中心点"。因此，相关的措辞要十分谨慎。如果中心点表达准确，读者马上就能了解论文的思想；并且，读者在阅读过程中盲目性就会减少，对实验意义的理解就会更好。

因为问题/目的是对整篇文章的概览，并且论文中的每个段落和句子都与之相关，因此，建议作者在撰写初稿前将研究问题/目的抄在便利贴上，并将这个便利贴贴在电脑显示屏的侧面和上端，以确保不会被忽视。这有助于提醒自己在论文

写作时专注于问题/目的。

论文的研究问题/目的应该指出研究的对象和特点。研究问题/目的通常不是以问题形式表述，而是用现在时以不定式短语或句子的形式表述。例如：To determine if…; Here we asked how …; In this study, we show that…; Here we examine the effects of…。

研究问题在逻辑上应该基于前文的陈述：哪些是已知的，哪些仍然未知或有待研究。研究问题的主题应该针对相应的问题。同样重要的是，研究问题应该是读者阅读完相关的背景介绍或有待研究的主题后所期望的问题。

4) 实验方法

引言指南 7：简要地说明研究方法

在研究论文的引言部分还应该简要地指出实验方法。通常情况下，实验方法只需一句话，最多二三句。实验方法在措辞上应该使读者能够快速识别。

例：实验方法的措辞

a. We analyzed X by agarose gel electrophoresis.

b. We simulated Tropical Instability Waves using a constant coefficient Laplacian friction scheme.

c. The structures of the compounds were characterized by UV, IR, 1H, HMR, 19FNMR spectra, and HRMS.

5) 结果与结论

在实验方法之后可简要提及结果和结论。尽管这方面的内容不是必需的，但读者通常希望通过阅读引言中就能了解结果和结论。大多数读者不喜欢非得读完全文才能找到问题答案。有些期刊要求将结果和结论放在论文的讨论部分。然而，如果有机会在引言中包括结果和结论，就尽量这样做。这会受到读者的欢迎。

可考虑指出研究发现为什么重要。通过以下例句的形式指出研究的重要性或意义是什么。

例：指出意义或重要性

a. X is an important addition to…

b. …which aids in the elucidation of…

如果在引言的最后指出重要性和意义，不仅可以以收敛的形式结束引言，而且可以向读者展示研究工作的总体前景。

4. 引言撰写的重要原则

1) 时态

在引言中通常是一个句子中有多种时态混用。报道已完成的行为，如引用先前的研究时，使用一般过去时。对于目前仍然有效的一般性陈述或事实性信息，即：研究发现是一般性的原理，可考虑使用现在时。说明研究问题/目的时，也可使用一般现在时，但在描述实验方法时则使用过去时态。应注意，引言中提到的过去已建立的知识应该采用一般现在时。如果采用过去时描述前人的工作，则暗示作者不认为这些结果是"事实"，只是"观察"而已。

2) 使用强有力的动词和短句子

与论文的其他部分相比较，引言的开始部分通常使用更宽泛的术语。因为引言的开始部分也更可能被相关同行阅读，因此，使用短句子和强有力的动词代替名词化的术语就显得尤其重要。

3) 连贯和衔接

引言指南 8：确保好的连贯和衔接

通常情况下，引言的所有部分都应有引导信号，以免读者难以判断作者所提供的相关信息。这些信号的变化取决于对已知、未知、问题和实验方法如何措辞，并且可能有大量的变化形式。

表 1　引言中的信号

背景	未知	问题、目的或发现	实验方法	结果	意义
X is…	..is unknown	We hypothesized that…	To test this hypothesis, we…	We found…	…consistent with…
X affects…	…has not been determined	To determine…	We…	…was found	…indicating that…
X is a component of Y	The question remains whether…	To study… To examine… To assess… To analyze… In this study we examined…	We analyzed… For this purpose, we…	We determined…	…make it possible to…
X is observed when Y happens…	…is unclear	Here we describe…	…by/using	Our findings were…	…may be used to…
X is considered to be…		Here we report… This report describes…	For this study we…	We observed that…	…is important for…
X causes Y	…does not exist …is not known	We examined whether X is… We assessed if… We determined if… We analyzed Y…	To evaluate… we… To answer this question, we…	Based on our observations…	Our analysis implies/suggests … Our findings indicate that…

5. 引言中的常见问题

引言中最常见的问题包括：

- 缺少某些元素(未知/问题，设问，实验方法)
- 某些元素的内容模糊
- 冗长
- 语境/背景过于狭窄

- 综述性句子

1) 引言中某些元素的缺失或模糊

如果引言中一个或多个元素缺失或模糊，读者就会感到困惑和沮丧，因为他们感到没有明白论文所表达的思想。最常见的元素缺失是未知和实验方法。有时甚至是没有（清楚地）指出研究问题/目的。当读者未能读到有关未知/问题、问题/目的或实验方法的明确信号时，元素的模糊性通常就出现了。

2) 冗长

这个问题的出现是因为作者对主题进行综述，而不是严格地以漏斗形式从背景信息向主题相关的信息收敛。如果作者对主题进行综述，读者就不知道主题的焦点或论文的重点了。

3) 背景过于狭窄

如果引言中提供的背景信息过于狭窄，大多数读者将不会有兴趣阅读全文，因为他们觉得从一开始就已迷失；如果引言中一般性的背景信息缺失、过于抽象或专业性太强，文章也将会失去读者。有关主题的一般性介绍的缺失会导致读者因不了解背景信息而对全文产生不知所云的困惑。同时，短句子比长句子更有分量。以短句子开头的引言不仅会令大多数读者易于理解，而且也更加有趣并且更能吸引人们的注意力。

4) 综述性句子

有时，作者会将研究性论文的引言与综述性论文的引言相混淆。在这种混淆的引言中，大多数句子都是信息性的，但是最后一句话通常是对论文的概述，指出某主题将在本文中阐述或讨论。然而这些陈述在研究性论文的引言中是没用的，应加以避免。

Reference: Angelika H. Hofmann. 科技写作与交流-期刊论文、基金申请书及会议讲演[M]. 任胜利，莫京，安瑞，等，译. 北京：科学出版社. 2012.

Unit 9　Heavy Metal Pollution

Part A　Intensive Reading

Heavy Metals

Definitions

Heavy metals are generally defined as metals with relatively high densities, atomic weights, or atomic numbers. The criteria used, and whether metalloids are included, vary depending on the author and context. In metallurgy, for example, a heavy metal may be defined on the basis of density, whereas in physics the distinguishing criterion might be atomic number, while a chemist would likely be more concerned with chemical behaviour. More specific definitions have been published, but none of these have been widely accepted. The definitions surveyed in this article encompass up to 96 out of the 118 chemical elements; only mercury, lead and bismuth meet all of them. Despite this lack of agreement, the term (plural or singular) is widely used in science. A density of more than 5 g/cm^3 is sometimes quoted as a commonly used criterion and is used in the body of this article.

The earliest known metals—common metals such as iron, copper, and tin, and precious metals such as silver, gold, and platinum—are heavy metals. From 1809 onwards, light metals, such as magnesium, aluminium, and titanium, were discovered, as well as less well-known heavy metals including gallium, thallium, and hafnium.

Some heavy metals are either essential nutrients (typically iron, cobalt, and zinc), or relatively harmless (such as ruthenium, silver, and indium), but can be toxic in larger amounts or certain forms. Other heavy metals, such as cadmium, mercury, and lead, are highly poisonous. Potential sources of heavy metal poisoning include mining and industrial wastes, agricultural runoff, occupational exposure, and paints and treated timber.

Physical and chemical characterisations of heavy metals need to be treated with caution, as the metals involved are not always consistently defined. As well as being relatively dense, heavy metals tend to be less reactive than lighter metals and have much less soluble sulfides and hydroxides. While it is relatively easy to distinguish a heavy metal such as tungsten from a lighter metal such as sodium, a few heavy metals such as zinc, mercury, and lead have some of the characteristics of lighter metals, and lighter metals such as beryllium, scandium, and titanium have some of the characteristics of heavier metals.

Heavy metals are relatively scarce in the Earth's crust but are present in many aspects of modern life. They are used in, for example, golf clubs, cars, antiseptics, self-cleaning ovens, plastics, solar panels, mobile phones, and particle accelerators.

Biological Role

Trace amounts of some heavy metals, mostly in period 4, are required for certain biological processes. These are iron and copper (oxygen and electron transport), cobalt (complex syntheses and cell metabolism), zinc (hydroxylation), vanadium and manganese (enzyme regulation or functioning), chromium (glucose utilisation), nickel (cell growth), arsenic (metabolic growth in some animals and possibly in humans) and selenium (antioxidant functioning and hormone production). Periods 5 and 6 contain fewer essential heavy metals, consistent with the general pattern that heavier elements tend to be less abundant and that scarcer elements are less likely to be nutritionally essential. In period 5, molybdenum is required for the catalysis of redox reactions; cadmium is used by some marine diatoms for the same purpose; and tin may be required for growth in a few species. In period 6, tungsten is required by some bacteria for metabolic processes. An average 70 kg human body is about 0.01% heavy metals (~7 g, equivalent to the weight of two dried peas, with iron at 4 g, zinc at 2.5 g, and lead at 0.12 g), 2% light metals (~1.4 kg, the weight of a bottle of wine) and nearly 98% nonmetals (mostly water).

A deficiency of any of these periods 4~6 essential heavy metals may increase susceptibility to heavy metal poisoning. A few non-essential heavy metals have also been observed to have biological effects. Gallium, germanium (a metalloid), indium, and most lanthanides can stimulate metabolism, and titanium promotes growth in plants (though it is not always considered a heavy metal).

Environmental Heavy Metals

Heavy metals are often assumed to be highly toxic or damaging to the environment. Some are, while certain others are toxic only if taken in excess or encountered in certain forms.

Chromium, arsenic, cadmium, mercury, and lead have the greatest potential to cause harm on account of their extensive use, the toxicity of some of their combined or elemental forms, and their widespread distribution in the environment. Hexavalent chromium, for example, is highly toxic as are mercury vapour and many mercury compounds. These five elements have a strong affinity for sulfur; in the human body they usually bind, via thiol groups (-SH), to enzymes responsible for controlling the speed of metabolic reactions. The resulting sulfur-metal bonds inhibit the proper functioning of the enzymes involved; human health deteriorates, sometimes fatally. Chromium (in its hexavalent form) and arsenic are carcinogens; cadmium causes a degenerative bone disease; and mercury and lead damage the central nervous system.

Lead is the most prevalent heavy metal contaminant. Levels in the aquatic environments of industrialised societies have been estimated to be two to three times those of pre-industrial levels. As a component of tetraethyl lead, $(CH_3CH_2)_4Pb$, it was used extensively in gasoline during the 1930s~1970s. Although the use of leaded gasoline was largely phased out in North America by 1996, soils next to roads built before this time retain high lead concentrations. Later research demonstrated a statistically significant correlation between the usage rate of leaded gasoline and violent crime in the United States; taking into account a 22-year time lag (for the average age of violent criminals), the violent crime curve virtually tracked the lead exposure curve.

Other heavy metals noted for their potentially hazardous nature, usually as toxic environmental pollutants, include manganese (central nervous system damage); cobalt and nickel (carcinogens); copper, zinc, selenium and silver (endocrine disruption, congenital disorders, or general toxic effects in fish, plants, birds, or other aquatic organisms); tin, as organotin (central nervous system damage); antimony (a suspected carcinogen); and thallium (central nervous system damage).

Nutritionally Essential Heavy Metals

Heavy metals essential for life can be toxic if taken in excess; some have notably toxic forms. Vanadium pentoxide (V_2O_5) is carcinogenic in animals and, when inhaled, causes DNA damage. The purple permanganate ion MnO_4 is a liver and kidney poison. Ingesting more than 0.5 grams of iron can induce cardiac collapse; such overdoses most commonly occur in children and may result in death within 24 hours. Nickel carbonyl ($Ni_2(CO)_4$), at 30 parts per million, can cause respiratory failure, brain damage and death. Imbibing a gram or more of copper sulfate ($Cu(SO_4)_2$) can be fatal; survivors may be left with major organ damage. More than 5 milligrams of selenium is highly toxic; this is roughly ten times the 0.45 milligram recommended maximum daily intake; long-term poisoning can have paralytic effects.

Other Heavy Metals

A few other non-essential heavy metals have one or more toxic forms. Kidney failure and fatalities have been recorded arising from the ingestion of germanium dietary supplements (~15 to 300 g in total consumed over a period of two months to three years).

Exposure to osmium tetroxide (OsO_4) may cause permanent eye damage and can lead to respiratory failure and death. Indium salts are toxic if more than few milligrams are ingested and will affect the kidneys, liver, and heart. Cisplatin $(PtCl_2(NH_3)_2)$, which is an important drug used to kill cancer cells, is also a kidney and nerve poison. Bismuth compounds can cause liver damage if taken in excess; insoluble uranium compounds, as well as the dangerous radiation they emit, can cause permanent kidney damage.

Exposure Sources

Heavy metals can degrade air, water, and soil quality, and subsequently cause health issues in plants, animals, and people, when they become concentrated as a result of industrial activities. Common sources of heavy metals in this context include mining and industrial wastes; vehicle emissions; lead-acid batteries; fertilisers; paints; and treated timber; aging water supply infrastructure; and microplastics floating in the world's oceans. Recent examples of heavy metal contamination and health risks include the occurrence of Minamata disease in Japan (1932~1968); the Bento Rodrigues dam disaster in Brazil, and high levels of lead in drinking water supplied to the residents of Flint, Michigan, in the north-east of the United States.

Source: https://en.wikipedia.org/wiki/Heavy_metals

Words and Phrases

metalloid ['metlɔɪd]　*n.* 非金属；[材] 类金属；准金属；*adj.* 非金属的；类似金属性的；准金属的

metallurgy [mə'tælədʒi]　*n.* 冶金；冶金学；冶金术

indium ['ɪndɪəm]　*n.* [化学] 铟

beryllium [bə'rɪlɪəm]　*n.* [化学] 铍

scandium ['skændɪəm]　*n.* [化学] 钪

titanium [tɪ'teɪnɪəm]　*n.* [化学] 钛

crust [krʌst]　*n.* 地壳；外壳；面包皮；坚硬外皮；*vi.* 结硬皮；结成外壳；*vt.* 盖以硬皮；在…上结硬皮

glucose ['glu:kəʊs]　*n.* 葡萄糖；葡糖（等于 dextrose）

arsenic ['ɑ:snɪk]　*n.* 砷；砒霜；三氧化二砷；*adj.* 砷的；含砷的

hexavalent [ˌhɛk'sævələnt]　*adj.* [化学] 六价的（sexivalent）

degenerative [dɪ'dʒenərətɪv]　*adj.* 退化的；变质的；退步的

hydroxylation [haɪdrɒksɪ'leɪʃən]　*n.* [有化] 羟基化

endocrine ['endəʊkraɪn; -krɪn]　*n.* 内分泌；激素；内分泌物；*adj.* 内分泌的；激素的

antimony ['æntɪməni]　*n.* [化学] 锑

agricultural runoff　农田径流

treated timber　防腐处理木材；已处理木料

tetraethyl lead　四乙铅

Questions

1. What are heavy metals?
2. What are potential sources of heavy metal poisoning?
3. What heavy metals are essential for certain biological processes?
4. As we all know, heavy metals are widely used in our daily life. What are they used for in modern life?

Sentence-making

1. heavy metal, toxicity, exposure
2. soil, source, copper
3. water, mining, vanadium
4. agricultural runoff, endocrine, metabolism
5. precipitate, mercury, respiratory

Part B   Extensive Reading

Phytoremediation of Heavy Metals: A Green Technology

Heavy Metal Contamination

Since the beginning of the industrial revolution, heavy metal contamination of the biosphere has increased considerably and became a serious environmental concern. Totally 53 elements are classified as heavy metals. Their densities exceed 5 g/cm^3, and they are known as universal pollutants in industrial areas. Heavy metal is defined as any element with metallic characteristics, such as density, conductivity, stability as cations, and an atomic number greater than 20. Generally, depending on the type of element and its location, the concentration of metals in the soil ranges from traces levels to as high as 100, 000 mg/kg. Contamination by heavy metals can be considered as one of the most critical threats to soil and water resources as well as to human health.

Pollution can be defined as an accidental or deliberate contamination of the environment with waste generated by human activities. Sources of metal contamination include anthropogenic and geological activities. Natural sources of heavy metals contamination usually result from the weathering of mines, which are created anthropogenically by themselves. Industrial pollutants, smelting, mining, military activities, fuel production and agricultural chemicals are some of the anthropogenic activities that cause metal contamination. Environmental contamination by heavy metals as a result of industrial and mining activities became widespread in the late 19th and early 20th centuries. The application of phosphate fertilizers to the agricultural soil has led to increase in Cd, Cu, Zn and As. Indeed, the increasing demand for agricultural products has led to extensive cultivation in agricultural lands. However, the excessive use of these agro-chemicals creates environmental problems, such as accumulation of these chemical substances in the soil and plant uptake.

Heavy metal pollution is a crucial environmental concern throughout the world. Unlike organic matter, these metals cannot be altered by micro-organisms. The half-life of the toxic elements is more than 20 years. Heavy metals, including Cd, Cu, Cr, Zn, Ni and Pb as critical pollutants, have an adverse effect on the environment, specifically at high concentrations in areas with severe anthropogenic activities. Although they are natural components of the earth's crust, heavy metals' biochemical equivalence and geochemical cycles have changed noticeably due to human activities. According to the United States Environmental Action Group (USEAG), this environmental problem has threatened the health of more than 10 million people in many countries (Environmental News Service, 2006).

Remediation of Heavy Metals

Current conventional methods to remediate heavy metal-contaminated soil and water, such as *ex situ* excavation, landfill of the top contaminated soils, detoxification, and physico-chemical remediation, are expensive, time consuming, and labor exhaustive increase the mobilization of contaminants, and destroy the biotic structure of the soil. Plants are able to metabolize substances produced in natural ecosystems. Phytoremediation is an approach in which plants are applied to detoxify contaminated areas.

Phytoremediation is a promising new technology that uses plants to clean up contaminated areas. It is a low cost, long term, environmentally and aesthetically friendly method of immobilizing/stabilizing, degrading, transferring, removing, or detoxifying contaminants, including metals, pesticides, hydrocarbons, and chlorinated solvents. Over the past 2 decades, it has become a highly accepted means of detoxifying contaminated water and soil. Historically, phytoremediation has been considered a natural process, first identified and proved more than 300 years ago. The specific plant and wild species that are used in this technique are effective at accumulating increasing amounts of toxic heavy metals. These plants are known as accumulators. They accumulate heavy metals at higher concentrations ($\geqslant$ 100 times) above ground than do non-hyperaccumulators growing in the same conditions, without showing any observable symptoms in their tissues. Phytoremediation can be applied to detoxify areas with trivial pollution of metal, nutrients, organic matter, or contaminants.

Phytoremediation can be classified into different applications, such as phytofiltration or rhizofiltration, phytostabilization, phyto-volatilization, phytodegradation, and phyto-extraction.

- Phytofiltration or rhizofiltration is the removal by plant roots of contaminants in waste water, surface water, or extracted ground water.
- Phytostabilization is a simple, cost-effective, and less environmental invasive approach to stabilize and reduce the bioavailability of contaminants by using plants. In fact, this approach uses plant roots to restrict the mobility and bioavailability of contaminants in the soil. Plants can reduce the future adverse effects of pollutants in the environment by keeping them from entering the ground water or spreading in the air. This method is applicable when there is no prompt action to detoxify contaminated areas (for example, if a responsible company only exists for a short time, or if an area is not of high concern on a remediation agenda). In this approach, the chemical and biological characteristics of polluted soils are amended by increasing the organic matter content, cation exchange capacity (CEC), nutrient level, and biological actions.
- Phyto-volatilization is the use of green plants to extract volatile contaminants, such as Hg and Se, from polluted soils and to ascend them into the air from their foliage.
- Phytoextration is a phytoremediation technique that uses plants to remove heavy metals, such as Cd, from water, soil, and sediments. It is an ideal method for removing pollutants from soil without adversely affecting the soil's properties. Furthermore, in this approach, metals accumulated in harvestable parts of the plant can be simply restored from the ash that is produced after drying, ashing, and composting these harvestable parts. Phytoextraction has also been called phytomining or biomining. This technology is a more advanced form of phytoremediation, in which high-biomass crops grown in the contaminated soil are used to bioharvest and recover heavy metals. It can be applied in mineral industries to commercially produce metals by cropping.

Selection of Hyper-Accumulator of Plants

Successful phytoremediation requires recognition of suitable plant species to accumulate metals in toxic levels as well as creating high biomass. Generally, the ideal plants for phytoextraction should have high capacity to accumulate toxic levels of metals in their aerial parts (shoots), high growth rates, and tolerance to high salinity and high pH. Moreover, these plants must produce high dry biomass, simply grown and completely harvestable, and must uptake and translocate metals to aerial parts efficiently. Overall, it is recommended to use the native plant species that grow locally near the site. These species are less competitive under local conditions and will reduce the metal concentration to an acceptable level for normal plant growth.

Advantages and Disadvantages of Phytoremediation

Advantages: Phytoremediation is a low-cost and effective strategy to clean up contaminated soil without requiring high-cost tools and expert human resources. As a green technology, it is applicable for different kinds of organic and inorganic pollutants and provides aesthetic benefits to the environment by using trees and creating green areas, which is socially and psychologically beneficial for all. In addition, as a practical approach to decontaminating soil and water, residues can be reused with minimal harm to the environment. Furthermore, the expansion of contaminants to air and water is reduced by preventing leaching and soil erosion that may result from wind and water activity.

Disadvantages: Time is the most serious limitation of phytoremediation, because this approach may require several years for effective remediation. Moreover, preserving the vegetation in extensively contaminated areas is complicated, and human health could also be threatened by entering the pollutant into the food chain through animals feeding on the contaminated plants. This technology is not impressive when just a small part of the contaminant is bio-available for plants in the soil. Beside, it is limited to the low or mildly contaminated areas enclosed by the plant root district.

Source: Asrari E. Heavy metal contamination of water and soil: analysis, assessment, and remediation strategies[J]. Journal of Thoracic Oncology Official Publication of the International Association for the Study of Lung Cancer, 2014, 9(6):47-48.

Ahmadpour P, Ahmadpour F, Mahmud T M M, et al. Phytoremediation of heavy metals: A green technology[J]. African Journal of Biotechnology, 2012.

Words and Phrases

phytoremediation [faɪtəʊr'mi:dɪeɪʃn] 植物修复
[例句] This review provides an overview of the development and application of phytoremediation on the remediation of heavy metal contaminated soils and sites.
contamination[kənˌtæmɪ'neɪʃn] *n*. 污染，玷污；污染物

[例句] However, contamination can occur at any point along the food chain, from field to table.

conductivity [ˌkɒndʌk'tɪvəti]   *n.* 导电性；[物][生理] 传导性
[例句] The material's large surface area, flexibility, low density, and high conductivity also make it compelling for energy storage.

cation ['kætaɪən]   *n.* 阳离子
[例句] It comes from the fact that in an electrochemical cell, the cations are those that are reduced.

remediation [rɪmiːdɪ'eɪʃn]   *n.* 补救；矫正；补习
[例句] If you do find an infected machine, the jury is out about how best to do remediation.

detoxification [diːˌtɒksɪfɪ'keɪʃn]   *n.* 解毒；[生化] 解毒作用
[例句] Drink at least 2 litres of still mineral water throughout the day to aid detoxification.

metabolize [mə'tæbəlaɪz]   *vt.* 使新陈代谢；使变形；*vi.* 新陈代谢
[例句] According to the oil spill council, fish do not metabolize oil the way mammals do.

aesthetically [iːs'θetɪkli]   *adv.* 审美地，美学观点上地
[例句] For me, a successful design is one that is both aesthetically challenging and functionally perfect.

accumulator [ə'kjuːmjəleɪtə(r)]   *n.* [机] 蓄能器；[电] 蓄电池；累计期权
[例句] There are no accumulators in this room, just current transducers.

phytostabilization [faɪtəstəbəlaɪ'zeɪʃn]   *n.* 植物稳定化
[例句] Mycorrhizae can bind metals and limit their translocation to shoot, and provide a new way for phytostabilization and protect plants against the toxicity of heavy metals.

invasive [ɪn'veɪsɪv]   *adj.* 侵略性的；攻击性的
[例句] An invasive alien marine creature that can smother native species has been found in Strangford Lough.

bioavailability [ˌbaɪəʊəˌveɪlə'bɪləti]   *n.* 生物利用度；生物药效率
[例句] The lack of dependence of this effect of aspirin on its systemic bioavailability suggests that it is platelet-mediated.

foliage ['fəʊliɪdʒ]   *n.* 植物；叶子（总称）
[例句] Dark foliage clothes the hills.

leach [liːtʃ]   *n.* [矿业] 沥滤；*v.* [矿业][冶] 浸出；[土壤] 淋洗；滤取；滤去
[例句] Heavily acidic conditions also prompt the leaching of toxic metals into nearby bodies of water.

non-hyperaccumulator 非超积累植物
phytofiltration 植物过滤
rhizofiltration 根际过滤
phyto-volatilization 植物挥发
phytodegradation 植物降解
phytomining 植物采矿；植物冶金
the industrial revolution 工业革命
atomic number 原子序数
military activities 军事活动
phosphate fertilizer [肥料] 磷肥
agricultural soil 耕种土壤
extensive cultivation 粗放耕作
plant uptake 植物吸收
adverse effect 副作用

anthropogenic activities 人类活动
geochemical cycle 地球化学循环
ex situ 非原位；天然状态外
time consuming 耗费时间的；旷日持久的
long term 长期的
chlorinated solvent 氯化溶剂
cation exchange capacity 阳离子交换量
toxic level 毒性水平
aerial part 地上部分
high salinity 高盐度
dry biomass 干重
soil erosion 水土流失；[土壤] 土壤侵蚀

Questions

1. What are the natural and anthropogenic sources of heavy metal contamination?
2. What is the difference between heavy metal and organic matter?
3. What are the remediation methods for heavy metal pollution? Which method does the article mainly talk about?
4. What are the applications of phytoremediation?
5. Please talk about the advantages and disadvantages of phytoremediation in your own words.

Part C Dialogue

Dialogue 1: Micro robots may remove heavy metals from wastewater efficiently

Todd: Hey, Marianne. Did you listen the latest Scientific American?
Marianne: Of course, they proposed a new solution to strip heavy metals from wastewater—a fleet of microscopic, self-propelled, aquatic robots.
Todd: Yes, the size of each one was just 10 lined-up bacteria—so tiny that a billion will fit in a syringe. Each tube-shaped microbot is a sandwich of three materials. A graphene outer layer, which binds to heavy metals. A middle layer of nickel, which gives the bots magnetic polarity, so they can be pulled through wastewater with magnets. And

platinum inside—for propulsion. So interesting!

Marianne: Yes! Just add a bit of peroxide to the wastewater, and it'll react with the platinum to form water and oxygen bubbles, which propel the tubes along. In an hour, a swarm of 200,000 bots scavenged 80 percent of the lead from three millimeters of tainted water. And the researchers estimate that it costs only about five cents a liter to do so.

Todd: Oh, I can't explain it. It's just fantastic! Compared with the traditional removal methods, such as chemicals, filters, membranes, and electric current, it's cheaper and more effective.

Marianne: The researchers envision the bots as a portable solution for small companies—they'd treat their water onsite instead of carting it to a treatment plant. And after the robots do the rounds, the heavy metals can be stripped away.

Todd: Following this solution, companies can reuse the heavy metals, and, ultimately, keep them out of our waterways.

Marianne: That will be wonderful!

Todd: Can't agree more!

Reference: http://www.kekenet.com/broadcast/201605/440607. Shtml

Dialogue 2: About 12 million of grain is polluted by heavy metals in China every year

Li: Hey, dear. What are you doing?

Zhao: I am reading the *Global Times*.

Li: Is there anything interesting?

Zhao: Experts warn that soil pollution has reached alarming levels in China, and will cause enormous economic losses and affect the health of people for generations!

Li: Oh, as I know, arable land in our country is shrinking due to soil pollution and is bringing huge losses to a country already short of farmland.

Zhao: Yes, China has 22 percent of the world's population but only 9 percent of the planet's arable land, and 12 million tons of crops in China are polluted by soil contaminated with heavy metals annually, which would cause a yearly loss of 20 billion yuan!

Li: Terrible.

Zhao: Heavy metal pollution also poses a great threat to public health as pollutants, and severely affects the quality of crops. The Guangzhou Food and Drug Administration revealed that among the 18 batches of rice tested during random checks, eight were found to contain excessive amounts of heavy metals, which is known as a strong carcinogen and can cause pathological changes in kidney and other organs. Furthermore, according to latest figures from the Ministry of Environmental Protection and the Ministry of Land and Resources, about 16.1 percent of China's surveyed land is polluted by heavy metals including cadmium, arsenic, lead and mercury.

Li: Preserving protected arable land should be a priority, surveillance of land should be enhanced, and cleaning up should be done as soon as pollution is detected.

Zhao: I think so. Last year, the central government allocated 2.8 billion yuan for anti-pollution projects in 30 prefecturelevel cities, but actually it is far from enough.

Reference: http://www.kekenet.com/read/201612/481533.Shtml

Part D   Supplementary Vocabulary

化学元素周期表

第 01 号元素：氢 H，Hydrogen
第 02 号元素：氦 He，Helium
第 03 号元素：锂 Li，Lithium
第 04 号元素：铍 Be，Beryllium
第 05 号元素：硼 B，Boron
第 06 号元素：碳 C，Carbon
第 07 号元素：氮 N，Nitrogen
第 08 号元素：氧 O，Oxygen
第 09 号元素：氟 F，Fluorine
第 10 号元素：氖 Ne，Neon
第 11 号元素：钠 Na，Sodium
第 12 号元素：镁 Mg，Magnesium
第 13 号元素：铝 Al，Aluminium
第 14 号元素：硅 Si，Silicon
第 15 号元素：磷 P，Phosphorus
第 16 号元素：硫 S，Sulfur
第 17 号元素：氯 Cl，Chlorine
第 18 号元素：氩 Ar，Argon
第 19 号元素：钾 K，Potassium
第 20 号元素：钙 Ca，Calcium
第 21 号元素：钪 Sc，Scandium
第 22 号元素：钛 Ti，Titanium
第 23 号元素：钒 V，Vanadium
第 24 号元素：铬 Cr，Chromium
第 25 号元素：锰 Mn，Manganese
第 26 号元素：铁 Fe，Iron
第 27 号元素：钴 Co，Cobalt
第 28 号元素：镍 Ni，Nickel
第 29 号元素：铜 Cu，Copper
第 30 号元素：锌 Zn，Zinc
第 31 号元素：镓 Ga，Gallium
第 32 号元素：锗 Ge，Germanium
第 33 号元素：砷 As，Arsenic
第 34 号元素：硒 Se，Selenium
第 35 号元素：溴 Br，Bromine
第 36 号元素：氪 Kr，Krypton
第 37 号元素：铷 Rb，Rubidium
第 38 号元素：锶 Sr，Strontium
第 39 号元素：钇 Y，Yttrium
第 40 号元素：锆 Zr，Zirconium
第 41 号元素：铌 Nb，Niobium
第 42 号元素：钼 Mo，Molybdenum
第 43 号元素：锝 Tc，Technetium
第 44 号元素：钌 Ru，Ruthenium

第 45 号元素：铑 Rh, Rhodium
第 46 号元素：钯 Pd, Palladium
第 47 号元素：银 Ag, Silver
第 48 号元素：镉 Cd, Cadmium
第 49 号元素：铟 In, Indium
第 50 号元素：锡 Sn, Tin
第 51 号元素：锑 Sb, Antimony
第 52 号元素：碲 Te, Tellurium
第 53 号元素：碘 I, Iodine
第 54 号元素：氙 Xe, Xenon
第 55 号元素：铯 Cs, Cesium
第 56 号元素：钡 Ba, Barium
第 58 号元素：铈 Ce, Cerium
第 59 号元素：镨 Pr, Praseodymium
第 60 号元素：钕 Nd, Neodymium
第 61 号元素：钷 Pm, Promethium
第 62 号元素：钐 Sm, Samarium
第 63 号元素：铕 Eu, Europium
第 64 号元素：钆 Gd, Gadolinium
第 65 号元素：铽 Tb, Terbium
第 66 号元素：镝 Dy, Dysprosium
第 67 号元素：钬 Ho, Holmium
第 68 号元素：铒 Er, Erbium
第 69 号元素：铥 Tm, Thulium
第 70 号元素：镱 Yb, Ytterbium
第 71 号元素：镥 Lu, Lutecium
第 72 号元素：铪 Hf, Hafnium
第 73 号元素：钽 Ta, Tantalum
第 74 号元素：钨 W, Tungsten
第 75 号元素：镧 La, Lanthanum
第 75 号元素：铼 Re, Rhenium
第 76 号元素：锇 Os, Osmium
第 77 号元素：铱 Ir, Iridium
第 78 号元素：铂 Pt, Platinum
第 79 号元素：金 Au, Gold
第 80 号元素：汞 Hg, Mercury
第 81 号元素：铊 Tl, Thallium
第 82 号元素：铅 Pb, Lead
第 83 号元素：铋 Bi, Bismuth
第 84 号元素：钋 Po, Polonium
第 85 号元素：砹 At, Astatine
第 86 号元素：氡 Rn, Radon
第 87 号元素：钫 Fr, Francium
第 88 号元素：镭 Ra, Radium
第 89 号元素：锕 Ac, Actinium
第 90 号元素：钍 Th, Thorium
第 91 号元素：镤 Pa, Protactinium
第 92 号元素：铀 U, Uranium
第 93 号元素：镎 Np, Neptunium
第 94 号元素：钚 Pu, Plutonium
第 95 号元素：镅 Am, Americium
第 96 号元素：锔 Cm, Curium
第 97 号元素：锫 Bk, Berkelium
第 98 号元素：锎 Cf, Californium
第 99 号元素：锿 Es, Einsteinium
第 100 号元素：镄 Fm, Fermium
第 101 号元素：钔 Md, Mendelevium
第 102 号元素：锘 No, Nobelium
第 103 号元素：铹 Lw, Lawrencium
第 104 号元素：𬬻 Rf, Rutherfordium
第 105 号元素：𬭊 Db, Dubnium
第 106 号元素：𬭳 Sg, Seaborgium
第 107 号元素：𬭛 Bh, Bohrium
第 108 号元素：𬭶 Hs, Hassium
第 109 号元素：䥑 Mt, Meitnerium
第 110 号元素：𫟼 Ds, Darmstadtium
第 111 号元素：𬬭 Rg, Roentgenium
第 112 号元素：鎶 Cn, Copernicium

Source: https://wenku.baidu.com/view/e371b1d533d4b14e84246805

Part E Supplementary Knowledge

如何写英文学术论文的材料与方法

1. 内容

材料与方法指南 1：提供足够的细节和参考文献，以便有经验的科学家能评估或重复作者的工作

材料与方法指南 2：描述材料和方法，而不是结果

1) **材料与方法**

材料与方法部分应包括：
- 材料(药物、培养基、缓冲液、气体或使用的器具等)
- 对象(患者、实验材料、动物、微生物、植物、土壤等)
- 设计(包括独立变量和因变量、实验组和对照组)
- 程序(什么、如何，以及为什么做)

应尽可能地阐述材料与方法，并且不能忘记对照实验。查阅拟投稿期刊"作者须知"中相关规定，并严格遵循。应注意，材料与方法部分不可避免地与结果部分有联系。在材料与方法部分，只需要描述如何获得结果。相反，在结果部分，则需要展示材料与方法部分所描述的所有结果。不要在材料与方法章节中混合报道结果，除非为阐述下一个实验步骤而必须提供的中间性结果。

2) **详情及技术规范**

在材料与方法部分，需要提供足够的细节和精确的技术规格，如温度、pH 值、总量、时间等数值，以确保实验能够被重复。如果必要，还应包括商标名称、制造商、型号等信息，此外，要指出生物组织的分类名称。

例：提供足够的细节

To identify genes with a high probability of having differential expression in adenomas and follicular carcinomas, we used statistical methods.

改后：

To identify genes with a high probability of having differential expression in adenomas and follicular carcinomas, **we used parametric (T test) and nonparametric (Mann-Whitney U test) methods.**

例：材料和方法中不必要的信息

Cells were scraped out of the wells and resuspended in a 1.5 mL Eppendorf tube.

在这个例子中，有关于试管尺寸的描述"in a 1.5 mL Eppendorf tube."就某些实验而言，如果对于实验的成功至关重要，提及某些设备的制造商可能很重要，但通常情况下试管尺寸大小的描述是不必要的细节，因此应该避免。

改后：Cells were scraped out of the wells and resuspended **in 100 μL sterile saline.**

3) **参考文献**

如果研究方法从没有报道过，应提供所有必要的细节；如果曾经在公开发表的期刊中报道过，只提供参考文献即可。

例：参考以前报道过的方法

Plasmids were isolated according to Braun (1).

如果是对以前发表的方法进行了改进，应提供文献资料，并给出详细的改进说明。

例：参考改进的以前报道过的方法

Plasmids were isolated according to Braun (1) with minor modifications. Instead of dissolving DNA pellets in sterile water, pellets were dissolved in buffer A.

一定要引用原始文献，即：引用实际提供实验方法的文献，而不是转引该方法的文献。

4) **附录**

材料与方法指南 3：将有关实验过程的详细描述或其他冗长的细节作为附件处理

有关实验过程的详细描述或其他冗长的细节最后不要作为论文的正文，而作为附件处理。这项工作应该是论文计划的一部分，所以附件也应与论文一起送交同行评议，而不是等到与校样一起才提交给期刊。或者，也可以将其作为补充材料发送到期刊所推荐的仓储存档。

除了实验过程的详细描述，包含在附录而不是研究论文(或建议书)正文中的材料可能还包括详细的计算、算法、证明、表格、线图和图像，或用于综合分析和比较的大型数据集。目前许多期刊都有保存补充材料(包括原始数据)的电子档案。这种安排使得作者在论文正文的布局上既全面，又简洁。

2. **组织**

　　材料与方法指南 4：实验细节的安排遵循时间顺序或小节原则
　　材料与方法指南 5：提示并关联不同的主题
　　材料与方法指南 6：阐述作用和功能不清的实验环节的目的

材料和方法通常是一个较长的章节并且常涉及各种主题。这些主题的组织和方法的描述(包括每个方法的过程)需要遵循一定的逻辑顺序。

在对材料和方法部分进行组织时可以将各实验阶段组合为一个或多个段落。信息类型一致的段落组合为小节。各小节的实验描述通常按时间顺序或重要程度展开。每个小节都有副标题，其作用相当于一个信息提示，表示特定的材料、变量、或特定的程序。虽然可选择次级小节的形式，但通常基于简化且清楚的原则。次级小节可以包含一个或多个段落。查阅拟投稿的期刊，了解一下相关领域期刊习惯使用何种次级标题，并调整自己论文的相关小节。甚至可以考虑在材料与方法及结果中使用同样的次级标题。

在任何情况下都要确保读者理解每个实验环节的实施目的及其与论文中心问题的相关性。因此，应该指出那些作用或关系与论文中心问题不甚清楚的实验环节的目的或理由。此外，要提供理解实验所必需的背景信息。有关目的或背景的阐述通常在段落的开头，并且通常发挥着主题句和过渡词的作用。

例：主题句/目的陈述

a. To purify prolyl 4-hyderoxylase from human placenta, full-term human placentae were collected 30 min after delivery.

b. Next, a trait-by-trait correlation matrix was developed to assist in identifying clusters of related traits.

3. **材料与方法的重要写作原则**

> **材料与方法指南 7**：材料和方法部分通常优先使用被动语态
> **材料与方法指南 8**：没有明显的理由时不要切换（语态）角度

1) 语态

通常，研究论文的材料与方法部分使用被动语态优先于主动语态。原因有二：一是可以将材料或方法作为句子的主题加以强调，二是读者不需要知道谁是实验的实施者。

例：语态的使用

The principal investigator collected the different fungal species from various tepuis in Venezuela.

改后：

Different fungal species were collected from various tepuis in Venezuela.

从某个角度来撰写整个材料和方法部分最为容易。缺点是如果大多数句子都是被动语态，写作会很枯燥。有经验的作者在写作时会考虑到单词的位置和关联性，会在材料和方法部分同时使用主动语态和被动语态，以使文字更加流畅、有趣和清晰。

不过，在同一个段落内要避免没有理由地进行主动和被动语态的切换。这种切换会不必要地混淆和分散读者的注意力。

例：语态的使用

The assays were performed for 10 min at room temperature. We then added 10 mL of 95% ethanol.

在这个例子中，作者没有理由地将被动语态换为主动语态。这类转换在非英语母语的作者中尤其常见。改后的例子一致性地使用了被动语态，读者更容易阅读。

改后：

The assays **were performed** for 10 min at room temperature. Then 10 ml of 95% ethanol **was added**.

2) 时态

在材料和方法部分，叙述已完成的动作使用过去时态。但是，表述一般有效的观点和目前仍然真实的信息，或解释图表的内容时应使用一般现在时。

例：时态的使用

a. Because mud volcanoes **emit** incombustible gases such as helium in close proximity to lava volcanoes, we collected gaseous samples from Lusi.

b. Criteria used in selecting subjects **are listed** in Table 2.

在许多描述性论文尤其是计算生物学论文中，方法部分使用一般现在时。

3) 单词的选择

材料与方法指南 9：仔细地选择词汇

在材料和方法中，需要描述精确而特定的事项，因此要避免行话和冗余。有些行话术语的使用频率很高以至于某些非英语母语者认为是"标准"的英语语法。这种例子包括用"bugs"代替"bacteria"（细菌），用"overnext"代替"the one after next"（下一个），用"western blotting"代替"western blot analysis"（蛋白质印迹分析）。非英语母语的作者尤其要注意，最好请英语母语的科学家或科学编辑帮助修改文字。

在选择使用某些特定单词时必须精确。如 determine，measure，calculate，quantitate 及 quantify 的区分。

determine	通过调查、计算、实验、鉴定或研究从而发现(to find by investigation, calculation, experimentation, survey, or study)
measure	测度大小、长度、数量、程度等(to find the size, length, amount, degree, etc.)
calculate	通过使用数字计算来解决或发现(to work out or find out something by using numbers, to compute)
quantitate	精确测量某物(to measure something precisely)
quantify	衡量某物的数量(to measure the quantity of something)

4. 道德行为

材料与方法指南 10：遵循道德行为准则

科研人员应该了解本国和合作国对有关研究对象，如人或动物的伦理、法律和法规的要求。在撰写论文前要核查这些要求。大多数期刊要求作者投稿时报告有关人或动物的研究是否符合相关政策和强制性要求。期刊编辑可能要求作者出示研究伦理委员会的书面批准书。

5. 材料与方法中常见的问题

研究论文中材料和方法部分最常见的问题：
(1) 细节信息缺乏；
(2) 实验目的缺失；
(3) 被动语态变为主动语态突兀；
(4) 过去时和现在时的切换缺乏理由。

Reference: Angelika H. Hofmann. 科技写作与交流-期刊论文、基金申请书及会议讲演[M]. 任胜利，莫京，安瑞，等，译. 北京：科学出版社，2012.

Unit 10 Environmental Biology

Part A Intensive Reading

Microbial Composition and Stoichiometry

Elemental Makeup

Most microorganisms are 70% to 90% water on a mass basis. The remaining dry weight is typically about 15% ash (minerals that remain upon combustion) and 85% volatile (mainly organic) material. The elemental composition of the dry matter of typical bacteria such as *Escherichia coli* is shown in Table 1. Of course, these values will vary among different strains and will also depend on the physiological state of the cell. It is sometimes useful to write an apparent chemical formula for microorganisms. Commonly, $C_5H_7O_2N$ has been used for this purpose (Table 2). Note that this formula gives reasonably good agreement with the values from *E. coli* for C, H, O, and N, which make up 92% of the total dry mass, but totally ignores the other elements. The elemental composition of some important cell constituents, including some storage materials, is shown in Table 2. Thus, although $C_5H_7O_2N$ is a useful simplification, it is not a true chemical formula nor an exact stoichiometric expression.

Table 1 Elemental Composition of a Microbial Cell

Element	Symbol	Atomic Weight	Cell Dry Weight (%)[a]	Element Ratio[b]	Formula[c]	Weight (%)
Carbon	C	12.01	50	4.2	5	53.1
Hydrogen	H	1.00	8	8.0	7	6.2
Oxygen	O	16.00	20	1.3	2	28.3
Nitrogen	N	14.01	14	1.0	1	12.4
Phosphoru	P	30.97	3	0.097		
Sulfur	S	32.07	1	0.031		
Potassium	K	39.10	1	0.026		
Calcium	Ca	40.08	0.5	0.012		
Magnesiu	Mg	24.30	0.5	0.021		
Iron	Fe	55.85	0.5	0.0036		
other			~1.8			

Notes: a. Based on *E. coli*.

b. Apparent stoichiometric formula of *E. coli* based on cell dry weight.

c. Useful stoichiometric ratio often used to write the components of a cell as a chemical compound formula.

Table 2 Elemental Composition (Mass %) of Some Important Microbial Cell Components

Name	Stoichiometric Formula	C	H	O	N	P	S
Glucose	$C_6H_{12}O_6$	40	6.7	53.3			
Cellulose, Starch, Glycogen	$(C_6H_{10}O_5)_n$	44.4	6.2	49.3			
Chitin	$(C_8H_{13}O_5N)_n$	47.3	6.5	39.4	6.9		
Protein[a]	$(C_{5.35}H_{7.85}O_{1.45}N_{1.45}S_{0.1})_n$	54.0	6.7	17.1			2.7
DNA[b]	$(C_{9.75}H_{12}O_6N_{3.5}P_1)_n$	38.3	4.1	31.4	16.1	10.1	
PHB	$(C_4H_6O_2)_n$	55.8	7	37.2			
Palmitic acid[c]	$C_{16}H_{32}O_2$	74.9	12.6	12.5			

Notes: a. Assuming about equal prevalence of all the amino acids.
    b. Assuming 50% G+C content (equal prevalence of all four bases).
    c. A common fatty acid.

The cell's requirements for C, O, and H are typically supplied by some combination of organic material, carbon dioxide, elemental oxygen, and water (or occasionally, hydrogen sulfide or methane). The other requirements can be loosely categorized as macronutrients, micronutrients, and trace elements, although the boundaries between these groups are not applied uniformly.

For microorganisms, N and P are typically considered as macronutrients (for plants, K would be added). These are needed in a mass ratio of about 5:1. The required C/N/P ratio, as a rule of thumb, is commonly said to be 100:5:1. However, in this case, much of the carbon is used as an energy source rather than to make cell constituents.

The term micronutrients usually include S and Fe, and probably K, Ca, and Mg. Trace nutrients would include the many other elements, such as cobalt (Co), nickel (Ni), copper (Cu), and zinc (Zn), needed in only very small amounts, usually for specific enzymes. Roles that a number of elements play in cell metabolism are summarized in Table 3.

Some organisms may need fairly high concentrations of another element for a special purpose. Diatoms, for example, need substantial amounts of silicon (Si) to construct their silica shells, and many testate amoebas need calcium for theirs. Other organisms may require small amounts of other elements. Molybdenum (Mo), for example, is needed in trace amounts by the nitrifying bacterium *Nitrobacter* to oxidize nitrite.

Other elements, such as sodium and chloride, may be present in fairly high concentrations within a cell. However, even if they are required for survival (e.g., for osmoregulation), they usually are not referred to as nutrients.

Table 3 Roles of Various Elements within Microorganisms

Element	Symbol	Important Cellular Roles
Carbon, hydrogen, oxygen	C, H, O	Major constituents of organic matter
Nitrogen	N	Proteins; nucleic acids; peptidoglycan
Phosphorus	P	Nucleic acids; membrane phospholipids; coenzymes; energy utilization (phosphorylation and ATP); present as phosphate (PO_4^{3-})
Sulfur	S	Amino acids cysteine and methionine, which give proteins much of their three-dimensional structure; coenzymes (including CoA)
Iron	Fe	Cytochromes and other heme and nonheme proteins; enzyme cofactor
Potassium	K	Major inorganic ion (K^+) in all cells; enzyme cofactor
Calcium	Ca	Major divalent ion (Ca^{2+}); enzyme cofactor; endospores
Magnesium	Mg	Major divalent ion (Mg^{2+}); enzyme cofactor; active in substrate binding; chlorophyll
Cobalt	Co	Coenzyme (vitamin) B_{12}
Copper	Cu	Specialized enzymes, including cytochrome oxidase and oxygenases
Manganese	Mn	Specialized enzymes, including superoxide dismutase; enzyme cofactor
Molybdenum	Mo	Specialized nitrogen enzymes (nitrate reductase, nitrogenase, nitrite oxidase) and some dehydrogenases
Nickel	Ni	Urease; required for autotrophic growth of hydrogen oxidizers
Selenium	Se	Specialized enzymes, including glycine reductase and formate dehydrogenase
Tungsten	W	Some formate dehydrogenases
Vanadium	V	Some nitrogenase enzymes
Zinc	Zn	Specialized enzymes, including RNA and DNA polymerases; enzyme cofactor
Silica	Si	Cell walls of diatoms (algae)
Sodium, Chlorine	Na, Cl	Transport processes; osmoregulation; required by halophilic bacteria

Growth Factors

Many microorganisms are able to grow in a system with a single organic carbon and energy source, such as a sugar, and inorganic forms of all other nutrients (e.g., N as ammonium). This means that they are able to synthesize all of the other organic molecules that they require. Photo- and chemoautotrophs, in fact, may be able to grow without the need for any organic substances at all, since they get their carbon from CO_2. However, other microbes may require a few or many specific essential organic molecules that they are unable to synthesize. Referred to as growth factors, they usually

fall into one of three categories:
- Vitamins, which are typically components of certain coenzymes
- Amino acids, the building blocks of proteins
- Purines and pyrimidines, the nitrogen-containing bases of nucleic acids

In some cases the growth factor required may depend on the other compounds present. The filamentous bacterium Sphaerotilus natans, for example, can grow with ammonium as the only nitrogen source if vitamin B_{12} (cyanocobalamin) is present, but otherwise requires methionine (a sulfur-containing amino acid).

In growing microorganisms in the laboratory, the growth medium (plural, media) is considered defined if its exact chemical composition is known. Growth factors can be added individually as part of a defined medium when required, but often a preparation such as yeast extract, made from natural products and containing many compounds of unknown composition, will be used instead. Such "undefined" media are referred to as complex.

Molecular Makeup

Most of the dry mass of cells is composed of macromolecules. Proteins typically account for over half of the total. Nucleic acids, polysaccharides, and lipids are also major components (Table 4).

Table 4 Typical Molecular Composition of Bacteria

	Cell Fraction (%) dry weight	Typical (Approx.) Molecular Weight (g/mol)	Cellular Role
Proteins	52	10^5	Structure and enzymes
RNA	16	$10^5 \sim 10^6$	Genetic
DNA	3	10^9	Genetic
Polysaccharides	17	$10^3 \sim 10^6$	Structure, genetic, storage
Lipids	9	10^3	Structure, storage
Inorganics and small organics	3	10^2	Enzyme cofactors, osmoregulation

Source: Vaccari D A, Strom P F, Alleman J E. Environmental Biology for Engineers and Scientists[M]. Hoboken :John Wiley & Sons, Inc., 2006.

Words and Phrases

stoicheiometry [stɔɪkʌɪˈɒmɪtri]
n. 化学计量
escherichia coli　大肠杆菌

amino acid　氨基酸
hydrogen sulfide　硫化氢
trace element　微量元素

filamentous bacterium 丝状细菌
sphaerotilus natans 球衣菌
combustion [kəm'bʌstʃən] *n.* 燃烧
volatile ['vɒlətaɪl] *adj.* 挥发性的
glucose ['gluːkəus] *n.* 葡萄糖
cellulose ['seljuləus] *n.* 纤维素
starch [stɑːtʃ] *n.* 淀粉；*vt.* 给…上浆
glycogen ['glɪkəudʒen] *n.* 糖原
chitin ['kaɪtɪn] *n.* 甲壳素
methane ['miːθeɪn; 'meθeɪn] *n.* 甲烷
macronutrient [ˌmakrəu'njuːtrɪənt] *n.* 大量元素
micronutrient [ˌmaɪkrəu'njuːtrɪənt] *n.* 微量营养素
diatom ['daɪətəm] *n.* 硅藻
nitrify ['naɪtrɪfaɪ] *vt.* 硝化；使与氮化合；用氮饱和
nitrobacter [naɪt'rɒbæktə] *n.* 硝化菌属
nitrite ['naɪtraɪt] *n.* 亚硝酸盐
osmoregulation [ˌɒzmə(ʊ)rɛgjuˈleɪʃn] *n.* 渗透调节
methionine [meˈθaɪəniːn] *n.* 蛋氨酸
synthesize ['sɪnθəsaɪz] *v.* 合成
purine ['pjʊəriːn] *n.* 嘌呤
pyrimidine [pɪ'rɪmɪdiːn] *n.* 嘧啶
yeast [jiːst] *n.* 酵母
macromolecule [ˌmɑkrəu'mɒlɪkjuːl] *n.* 大分子
polysaccharide [ˌpɒlɪ'sækəraɪd] *n.* 多糖
lipid ['lɪpɪd] *n.* [生化] 脂质；油脂

Questions

1. What is the composition of microbes?
2. What are the categories of growth factors?
3. What is the definition of stoichiometric ratio?
4. Please briefly explain the role of macronutrient elements in microbes.
5. What is the cellular role of typical molecular composition of bacteria?

Sentence-making

1. methane, synthesize, microbe
2. macronutrient, micronutrient, trace elements
3. carbon, nitrogen, hydrogen
4. energy, nutrients, microorganism

Part B Extensive Reading

Soil and Groundwater Treatment-Phytoremediation

Phytoremediation is the use of plants to remove and/or biotransform contaminants.

The process of phytoremediation is comparable to that of constructed wetlands. Both applications make use of macro-and microscale biology, and both concepts have captured considerable attention, even though they each still qualify as evolving technologies given their relatively short histories. However, phytoremediation also has several important distinctions. First, the focus of phytoremediation is that of biologically remediating contaminated soils, sediments, and waters (both surface and ground water) as opposed to wetland applications focused solely on wastewater treatment. Second, phytoremediation systems may employ trees as well as smaller plants as the primary biological agent. Indeed, for those applications in which they are suitable, phytoremediation systems may offer a highly attractive "green" means of decontaminating lands and water. Successful applications have been achieved experimentally with a wide range of inorganic (e.g., metals, ammonia) and organic contaminants.

The potential benefits are again much the same as those of constructed wetlands, including that of an apparently simple, solar-driven, aesthetically pleasing, in situ "green" technology with few, if any, complex or energy-intensive hardware or operational requirements (i.e., as compared to conventional treatment operations that employ pumps, mixers, aerators, etc. that routinely use energy and require careful operator attention). These systems can also be self-sustaining in terms of procuring nutrients, they can make a beneficial contribution to the balance of water in their soils, they can establish a highly evolved complement of degradative enzymes, and they tend to be inexpensive both in their initial startup and in subsequent maintenance. Granted, this remediation approach will not work in all situations, and in even when it is successful, the remediation process will operate on a time-scale measured in years rather than in hours or days. The public perception of phytoremediation is extremely high, though, as a natural means of promoting the restoration of chemically contaminated sites.

However, the seemingly simplistic notion of using plants and trees to clean up these contaminated sites actually involves a far more sophisticated process than what is apparent to the eye. As was the case with constructed wetlands, the visibly "green" above ground portions of these systems are but a part, and in some cases perhaps even a lesser part, of an integrated remediation scheme that encompasses a complex array of physical, chemical, and certainly biological treatment factors. The type, density, and nurturing of the plants and trees is important as well as the nature of the soils (e.g., soil type, conductivity, depth to groundwater, nutrient availability) and climate (e.g., rainfall frequency and duration, radiation, seasonal climate, wind speed, humidity) in which they are grown. Finally, the character, concentration, location, and form (e.g., whether it is sorbed, soluble, solid) of the contaminating materials are also important factors.

The level of the groundwater table may vary considerably from one location to another, and also according to temporal changes in precipitation and climate, but in most instances it lies many meters below the surface and at a level not usually reached by plant root systems. As a result, phytoremediation was developed with systems whose remediating activity took place almost totally within the uppermost layer of the unsaturated zone (i.e., at shallow depths). However, subsequent developments with the nurturing and use of deep-rooting plants and trees have now expanded this technology down to the saturated region, at which point phytoremediation could then deal with contaminants extending fully down to the groundwater table.

Once the remediating plants and/or trees have been introduced successfully into a site, the means by which they can attempt to degrade or remove a group of involved contaminants can be quite diverse. An important, yet all too easily overlooked aspect of phytoremediation is that the plants are generally not the sole means of contaminant treatment. Granted, the plants themselves may contribute many different and important remediating effects (e.g., phytovolatilization), but in most instances their remediation role is metabolically complemented, perhaps even dominated by that of the microbes that are motivated correspondingly to live within the same soils. Figure 1 provides an overall synopsis of these prospective plant and microbial mechanisms.

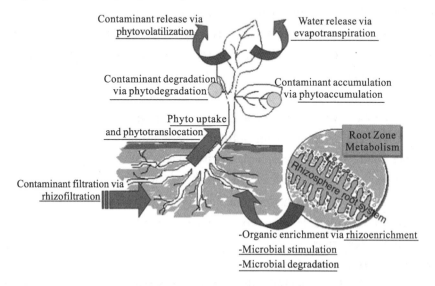

Figure 1 Phytoremediation mechanisms

The first such mechanism, rhizo-enrichment, stems from the fact that plants release into soils a number of exudates that are rich in organic carbon and that, in turn, effectively nurture the growth of many soil microorganisms. A sizable fraction of the carbon fixed through photosynthesis is released into soils, with estimates ranging from 10% to 30%. This material includes a range of readily biodegradable materials with small to moderately sized molecular weights, including sugar, protein, alcohol, and acids. Yet another group of organic carbon residuals are also released into soils by the senescence (aging) and decay of plant tissue, particularly that of fine-root biomass. There is also a beneficial physical impact with the growth and aging of plant roots, in that they tend to loosen the soil during both their growth and death, forming new paths for transporting water and aeration. This process subsequently tends to pull water to the topsoil surface while drying the lower saturated zones.

The latter enrichment of soils with organic carbon compounds exuded by plants subsequently promotes and maintains a significant enhancement in the growth of microbes within the immediate vicinity of the roots (i.e., microbial stimulation). There are actually two mechanisms by which plants provide this stimulation: by feeding the microorganisms with their exudates and by promoting the availability of oxygen. Here again, the photosynthetic activity of the plant is important, with at least some of its newly created oxygen being effectively pumped through the roots into the soil. Channeling created by roots, both alive and dead, also provides a means of physically opening the soil matrix and improving its porosity. The net effect of the added substrates and improved oxygen availability leads to levels of microbial activity and density that are considerably higher than those of barren, unvegetated soils, by several orders of magnitude.

This symbiotic relationship between plants and their adjacent microbial consortia stimulates microbes, which in return assist the plants in securing nutrients and essential vitamins. Given the diversity of the substrate forms available to these microbes and the variable nature of the rhizospheric environment (i.e., with dynamic changes in oxygen content, soil water presence, pH, etc.), a wide range of microbial types is found in these soils. In turn, the metabolic breadth of these bacterial and fungal forms encompasses a considerable range of enzymatic mechanisms and pathways. The resulting, collective effect of these microbes is that they can be expected, either directly or indirectly, to play a significant role in degrading organic contaminants present in the soils (i.e., microbial degradation). Compared to readily biodegradable compounds, recalcitrant organic contaminants found in soils may not be directly oxidized by these root-zone microorganisms as an energy source, but they may nonetheless be converted. Indeed, in the presence of other biodegradable root exudates, the catabolic enzymes generated to catalyze these reactions sometimes cooxidize the recalcitrant materials through a cometabolic conversion. The relative contribution of plants vs. microbes to degradation no doubt varies from one situation to the other, and there are those who would argue that one or the other typically plays a more dominant role. However, irrespective of which contribution might be dominant, the fact remains that the efficacy of phytoremediation commonly involves a coordinated and harmonious set of biological mechanisms that span the micro- to macroscale of life.

Contaminants not degraded by rhizospheric microbes are available for plant uptake by roots, where they may then either be retained or translocated farther upward into a plant's shoots and leaves (i.e., phyto uptake). Some plants simply uptake contaminants and store them in their roots, whereas others both uptake and translocate contaminants. There is, admittedly, a degree of uncertainty about the nature of these combined processes and the conditions under which they may each take place, but it appears that the polarity of the contaminants is an important factor.

Some plants not only transport contaminants across their cell membranes (i.e., through phyto uptake) but also move these materials internally beyond their roots (i.e., phytotranslocation). Here again, the polar vs. nonpolar nature of the contaminant is an issue, as well as the rate of transpiration being maintained by the plant. This rate of transpiration is, in fact, a key variable for translocation, with an apparent direct correlation between these two factors. Once a contaminant has been taken into a plant through these sequential processes of phyto uptake and phytotranslocation, this contaminant could theoretically then be removed from the site by harvesting and subsequent disposal of the plant's above- ground biomass.

Following uptake and translocation, many plants have evolved compound-specific detoxification pathways that involve subsequent conjugation and compartmentation reactions that effectively bind contaminants into their structural makeup (i.e., phytoaccumulation). These biotransformed and phytoaccumulated compounds can either be deposited into vacuoles or converted into insoluble (and frequently covalent) complexes within cell wall components through a process known as lignification. In some cases, the accumulated compounds are passed unchanged into these deposits; in other instances the material being accumulated is that of degradation fragments produced through preceding biochemical conversions that transformed the contaminants into nonphytotoxic metabolites. However, it is also possible that some plants may accumulate contaminants internally to a level where an ecotoxicological hazard develops that would severely restrict subsequent consumption or disposal of the plants.

Phytoremediation plants can also produce a number of enzymes that may promote the internal metabolism and degradation of contaminants (i.e., phytodegradation). For example, nitroreductase enzymes can initiate the breakdown of nitroaromatic munitions; dehalogenase enzymes will promote the degradation of chlorinated compounds; nitrilase will contribute to the degradation of herbicides; phosphatases will facilitate the catalysis of organophosphates; and peroxidases will promote the destruction of phenols.

The next process, phytovolatilization, theoretically involves the uptake and translocation of contaminants into leaves; plants may then release these compounds into the atmosphere through a volatilization mechanism. One particular plant, *Arabidopsis* (in the mustard family) has been found to produce a specific enzyme, mercury reductase, which reduces mercury to elemental mercury, which is then amenable to volatilization and release. Yet another known volatilization sequence involves the treatment of selenium-contaminated soils by rice, broccoli, and cabbage through the production of volatile dimethyl selenide and dimethyl diselenide. In addition, there are a number of low molecular weight VOC-type organic molecules that appear to be easily translocated and volatilized by various plants. The extent to which the latter reactions actually take place under real-world conditions, however, is not well established.

Although roots generally cannot be harvested in a natural environment, another phytoremediation process, rhizofiltration, can be used where plants are raised in greenhouses and transplanted to sites to filter metals from wastewaters biochemically. As the roots become saturated with metal contaminants, they can be harvested and disposed of. Phytoremediation has also been used in this fashion to concentrate radionuclides via rhizofiltration.

Extensive water uptake and release rates can also be maintained by a number of plants, including poplars, cottonwoods, and willows, in a fashion that will effectively pull contaminated groundwater plumes toward and through these phytoremediating tree roots (i.e., evapotranspiration). A single, mature willow tree, for example, can transpire more than 19 m^3 of water each day (~5000 gallons, or about 3.5 gal/min), and 1 ha (10,000 m^2) of a herbaceous plant such as saltwater cord grass has been found to evapotranspire even four times as much. There are several interrelated issues, including plant type, leaf area, nutrient availability, soil moisture, wind conditions, and relative humidity.

The principles and practice of phytoremediation systems involve several important engineering aspects, but in reality the procedures still qualify as an emerging technology. The issues that must be considered include those of the involved soil characteristics, the targeted contaminants and current concentrations, and the relative depth of the existing residuals.

Concerns regarding soil type stem from the fact that various plants have different preferences for either fine- or coarse-grained soils, which probably reflects the ability of the soils to hold and transfer varying amounts of moisture, air, and nutrients. The site-specific and perhaps seasonally fluctuating depth to the groundwater table is also important, as it affects the means by which a plant can draw water.

The majority of phytoremediation systems, such as those depicted schematically in Figure 2, apply to soil depths extending to the first 2 to 3 m. In turn, most of these plants probably draw their water either from roots closely aligned to the surface or water drawn from vadose-zone pore moisture. Most poplars, for instance, tend to have shallower root systems, and as a result, these types of plants have an inherent level of reliance on water being precipitated into, and then passed through, the surface soils.

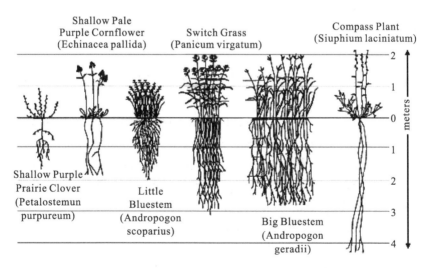

Figure 2 Representative phytoremediation plant variations

However, there are also deep-rooting plants that maintain a more water-loving life-style, in which their root systems extend into the capillary soil region or underlying saturated groundwater zone. These phreatophytic plants tend to have remarkable high summertime water uptake rates, possibly as a competitive means of trying to restrict the growth of their fellow plants. These deep-rooting plants generally have higher levels of plant biomass as well as higher overall growth rates.

When motivated to adopt this phreatophytic mode, though, alders, ash, aspen, river birch, and poplar have proven to be fast growers and rapid water users, with daily uptake rates during peak summertime periods ranging from 100 to as much as 1000 L/day per tree. The resulting uptake of water from the deep, saturated soil zone may actually produce a sizable depression of 5 to 10 cm in the water table within the capture zone where water is being used by the trees. Field studies of this sort have been able experimentally to develop hydraulic barrier strips using deeply rooted trees planted in rows aligned perpendicular to the direction of travel for the contaminated plume.

Of course, the actual process of evapotranspiration depends not only on the location and depth of the roots, but also on the number of plant and atmospheric parameters. The rate of water use by plants depends on the conductance of water through the plant stoma as well as the cumulative surface area of the leaves through which the water will finally be released into the atmosphere. The air temperature, wind speed, humidity, and radiation intensity will also play a part in the final rate of this release.

Source: Vaccari D A, Strom P F, Alleman J E. Environmental Biology for Engineers and Scientists[M]. Hoboken :John Wiley & Sons, Inc., 2006.

Words and Phrases

degradative [dɪˈgreɪdətɪv]　*adj*. 使下降（或降解、陵削、退化等）的，趋向下降（或降解等）的
[例句] Both in laboratory and field studies, attempts have been made to speed these biodegradation processes by adding known degradative microorganisms.
saturate [ˈsætʃəreɪt]　*vt*. 浸透；使饱和；*adj*. 浸透的，饱和的；深颜色的
[例句] If the filter has been saturated with motor oil, it should be discarded and replaced.
metabolic [ˌmetəˈbɒlɪk]　*adj*. 变化的；新陈代谢的
[例句] Regular exercise can increase your basal metabolic rate.
synopsis [sɪˈnɒpsɪs]　*n*. 概要，大纲
[例句] For each title there is a brief synopsis of the book.
exudate [ˈɛksjʊdeɪt]　*n*. 分泌液；流出物
[例句] The dormant spores of the fungus had the highest germination percentage (75%) in the root exudate of the susceptible host.
matrix [ˈmeɪtrɪks]　*n*. [数] 矩阵；模型；[生物] 基质；母体；子宫；[地质] 脉石
[例句] A mass of rock matrix which neither transmits nor stores water in significant

quantities is called an "aquifuge".
porosity [pɔː'rɒsəti] *n.* 有孔性，多孔性
[例句] Once the material solidifies, this effectively traps different porosity distributions throughout the solid matrix.
magnitude ['mægnɪtjuːd] *n.* 大小；量级；[地震] 震级；重要；光度
[例句] Even where models agree on the direction of rainfall change, there are considerable differences in magnitude.
symbiotic [sɪmbɪ'ɒtɪk] *adj.* [生态] 共生的
[例句] Corals are actually tiny organisms that join together in colonies, and they depend on a symbiotic relationship with certain species of algae to produce energy through photosynthesis.
rhizospheric [raɪ'zɒsferɪk] *adj.* 根际的
[例句] Rhizospheric nitrifying bacteria tended to increase under drought stress.
polarity [pə'lærəti] *n.* [物] 极性；[生] 反向性；对立；[数] 配极
[例句] The output is connected to a battery with incorrect polarity.
transpiration [ˌtrænspɪ'reɪʃn] *n.* [植] 蒸腾作用
[例句] Plants release water through their leaves by transpiration.
detoxification [diːˌtɒksɪfɪ'keɪʃn] *n.* 解毒；[生化] 解毒作用
[例句] Whilst the detoxification requires energy, slow decay may still occur.
compartmentation [kəmpɑːrtmɛn'teɪʃən] *n.* 分隔；区分
[例句] A complete analysis of the compartmentation of hydrolases is difficult to achieve.
vacuole ['vækjuəʊl] *n.* [细胞] 液泡
[例句] Vacuoles in cells appear to be hollow sacs but are actually filled with fluid and soluble molecules.
lignification [ˌlɪgnɪfɪ'keɪʃən] *n.* 木质化
[例句] This lignification serves to stiffen and strengthen the cells.
nitroreductase [naɪtrɔːrɪdʒʌk'tæs] *n.* 硝基还原酶
[例句] Compared with chemical reduction, it can make reacting condition mild and safe to use nitroreductase as catalyzer.
dehalogenase [dehələʊdʒe'neɪs] *n.* 脱卤素酶
[例句] Hydrolytic dehalogenation catalyzed by haloalkane dehalogenase is important to chemosynthesis and has quite potential use in environmental pollution.
nitrilase ['naɪtrɪleɪs] *n.* 腈水解酶
[例句] Using this method, 9 strains with relatively high nitrilase activity were obtained.
herbicide ['hɜːbɪsaɪd] *n.* [农药] 除草剂

[例句] One drills for oil and gas, the other distributes herbicides and pesticides.
organophosphate [ˌɔːɡənəʊ'fɒsfeɪt]   *n.* 有机磷酸酯（肥料）；*adj.* 有机磷酸酯的
[例句] The effects of organophosphate compounds on the organism were sum up.
peroxidase [pə'rɒksɪdeɪs]   *n.* [生化] 过氧化物酶
[例句] The impact of different degrees of degradation on activities of urease and peroxidase varied.
phenols ['fiːnɒlz]   *n.* 酚类
[例句] The major components of acidic fraction are phenols and organic acids.
volatilization [ˌvɒlətɪlaɪ'zeɪʃən]   *n.* 蒸发，挥发
[例句] Gallium and arsenic can be removed as halide by volatilization.
arabidopsis [əˌrabɪ'dɒpsɪs]   *n.* 拟南芥；阿布属
[例句] This work provides the basis for further molecular cloning and functional analysis of this gene in Arabidopsis pollen development.
mercury ['mɜːkjəri]   *n.* [化] 汞，水银
[例句] Mercury has a much greater density than water.
reductase [rɪ'dʌkteɪz]   *n.* [生化] 还原酶
[例句] This review mainly gives a brief account of the progress of research on aldose reductase inhibitors from microorganisms.
radionuclide ['reɪdɪəʊ'njuːklaɪd]   *n.* [核] 放射性核素
[例句] The severity of the risk depends on the radionuclide mix and the level of contaminant released.
moisture ['mɔɪstʃə(r)]   *n.* 水分；湿度
[例句] These plants imbibe moisture through their roots.
hydraulic [haɪ'drɔːlɪk]   *adj.* 液压的；水力的
[例句] The hydraulic pump is removed and installed in the assembly with the reservoir.
perpendicular [ˌpɜːpən'dɪkjələ(r)]   *n.* 垂线；*adj.* 垂直的
[例句] When a surface is perpendicular to the rays, their intensity is at its maximum.
parameter [pə'ræmɪtə(r)]   *n.* 参数；系数；参量
[例句] You need to set various parameters, such as source device and format.
stoma ['stəʊmə]   *n.* [植] 气孔；叶孔
[例句] Oxygen and carbon dioxide from the air enter through the stoma.
constructed wetland 人工湿地
saturated zone 饱和带；饱和层
dimethyl selenide 二甲基硒醚；二甲基硒

Questions

1. What is phytoremediation?

2. Please briefly describe the differences and linkages between phytoremediation and wetland treatment.

3. What is phytovolatilization?

4. What are the factors that affect plant evapotranspiration?

Part C   Dialogue

Dialogue 1: Marine microbes change their environment simply by defecation

Todd: Hey, Marianne. Did you read about the findings published in the Proceedings of the National Academy of Sciences this last weekend?

Marianne: Oh, you mean Dr. Bianchi's research?

Todd: Yes, Dr. Bianchi and his team tracked these ocean migrations with sonar data, and their findings are so interesting!

Marianne: I can't agree more! Every day, the largest mass migration on the planet happens in the world's oceans.

Todd: Tiny fish, jellies and shrimpy things feed at the water's surface by night. And by day, they hide in darker waters a few hundred meters below!

Marianne: Because the ocean is a dangerous place, and so swimming down to depth is your best bet to avoid predators.

Todd: And Dr. Bianchi found that the creatures descend to areas of deep water where certain species of bacteria hang out. Those bacteria snack on nutrients that float down from the surface—so-called marine snow.

Marianne: But the migrating creatures may also deliver food to the bacteria—in the form of ammonia in the creatures' urine. The bacteria metabolize the ammonia to produce energy and nitrogen gas—effectively removing the nitrogen from the food chain, and sending it in gaseous form back into the atmosphere. Then other bacteria fix that nitrogen gas back into food chains—on land and in the ocean—where it eventually finds its way into amino acids. Some of which make up the proteins in us. There are about 20 times more of these tiny fish than there are humans on the planet!

Todd: So these miniscule creatures could play an important role in the ocean's nitrogen cycle simply by taking a leak.

Marianne: That's amazing!

Reference: http://www.kekenet.com/broadcast/201501/355348.Shtml.

Dialogue 2: The researchers found organisms that inject methane live in extreme conditions

Marianne: Morning, Todd.
Todd: Morning, what are you doing, Marianne?
Marianne: I just finished listening to Scientific American—60-Second Science.
Todd: I like this program too. What interesting discovery was reported today?
Marianne: Researchers have found evidence of methane-producing life in some extreme conditions for the first time.
Todd: Some extreme conditions? I know the first life on Earth appeared about four billion years ago, and one place these pioneering organisms may have emerged is at hydrothermal vents, deep underwater.
Marianne: Yes, the unusual chemistry provided energy for primitive life-forms to survive, and life-forms like the methane-belching microbes were found at the vents today!
Todd: Interesting, where they found these methane-producing lives?
Marianne: At a spring in northern California, called The Cedars. The water there is extremely basic, with a pH of 11.6, and it contains no oxygen.
Todd: Not an easy place to survive.
Marianne: Yes, and researchers tested water and sediment at the Cedars. They found some samples got dosed with mercuric chloride to kill any life present. Those dosed samples produced no methane, but the samples in which microbes were allowed to survive did put out methane.
Todd: So at least some of the methane at the springs is indeed biological in origin. This finding may have implications for climate change alleviation. I remember that a geologically similar spring in Oman has been proposed as a site for carbon storage by pumping CO_2 underground, where it gets incorporated in stone.
Marianne: But the extremophiles at The Cedars can use CO_2 to make methane, an even more potent greenhouse gas. So imagine pumping CO_2 into the ground and having it come back up as methane.
Todd: Terrible, but this may not necessarily happen.
Marianne: However it is something to be tested for before fully implementing a carbon capture and storage technology at one of these types of sites.

Todd: You are right, besides, life is tenacious!

Marianne: So we should not let our biases prevent us from looking for evidence of life in what we would otherwise consider an unexpected place, including other planets and moons.

Reference: http://www.kekenet.com/broadcast/201606/450740-2.Shtml.

Part D   Supplementary Vocabulary

Environmental biology

activated biofilter   活性生物滤池
acute toxicity test   急性毒性实验
aerobic biological treatment   好氧生物处理
agricultural dreg   农业残渣
algae bloom   水华
anaerobic process   厌氧生物处理
autotrophy   自养生物
bioaccumulation   生物积累
bioaerosol   生物气溶胶
bioassay   生物检测
biocenological effect   生物群落效应
bioconcentration   生物浓缩
bio-diversity   生物多样性
biological degradation   生物退化
biological filter   生物滤池
biological flocculation   生物絮凝作用
biological fluidized bed   生物流化床
biological monitoring   生物监测
biological nitrogen removal   生物脱氮
biological rotating disc   生物转盘
biomagnification   生物放大
biomembrane; biological film   生物膜
bioseston   生物悬浮物
biotic nutrient   生物营养物质
biotreatment of gaseous pollutant   气体生物净化
biotreatment of odor   生物脱臭
chronic toxicity test   慢性毒性实验
commensalisms   偏利共生
crop rotation   轮作
derivative [dɪˈrɪvətɪv]   n. 衍生物
dilution [daɪˈluːʃn]   n. 稀释
fungicide [ˈfʌngɪsaɪd]   n. 杀菌剂
habitat [ˈhæbɪtæt]   n. 生境
herbicide [ˈhɜːbɪsaɪd]   n. 除草剂
heterotroph [ˈhetərətrəʊf]   n. 异养生物
indicator organism   指示生物
insecticide [ɪnˈsektɪsaɪd]   n. 杀虫剂
intercrop [ɪntəˈkrɒp]   n. 间作作物
interplant [ɪntəˈplɑːnt]   n. 套种
law of minimum   最小因子定律
life cycle   生命周期
limit of tolerance   耐受极限
marsh gas   沼气
mesosaprobic zone   中污生物带
microorganism intrinsic metabolism   微生物内源代谢
microorganism synthetic metabolism   微生物合成代谢
mutualism [ˈmjuːtʃʊəlɪz(ə)m]   n. 互利

共生
non-persistent pesticide 无残留农药
oligosaprobic zone 寡污生物带
open season 猎(渔)期
over-hunting; over-fishing 过度捕获
parasitism ['pɑrəsɪtɪz(ə)m] *n.* 寄生
phytopesticide 植物性农药
piled manure 堆肥
pollution indicating organism 污染指示生物
polysaprobic zone 多污生物带
preservative [prɪ'zɜːvətɪv] *n.* 防腐剂
primary pathogen 原生病原体
primary pollutant 原生污染物
purification pond 氧化塘/生物塘
red tide 赤潮
restoration ecology 恢复生态学
rodenticide *n.* 杀鼠剂
screening test; range-finding test;
preliminary test 预备实验
secondary pollutant 次生污染物
sensitive species 敏感种
sewage irrigation 污水灌溉
shelter belt 防护林带
sterilization [ˌsterəlaɪ'zeɪʃn] *n.* 灭菌
tolerant species 耐污种

Part E Supplementary Knowledge

如何写英文学术论文的结果

1. 概述

结果部分是作者研究的主要科学贡献。鉴于引言中已介绍了研究背景并指出了论文的研究目的/问题，与讨论部分联系，结果部分是论文的核心或骨架。

2. 内容

1) 一般性内容

　　结果指南1：报道主要结果以及其他重要的发现
　　结果指南2：向读者介绍图表中的数据
　　结果指南3：要包括对照控制组的结果

结果部分介绍实验结果，并向读者表明图表中数据的意义。报道的结果应该与引言中提及的信息，以及实验材料与方法的描述直接相关。应排除初步的结果和不相关的结果。不要忘了包含控制对照的结果，如果需要，简要介绍实验目的。此外，必要时还应一并介绍是否支持假设的结果，并说明任何矛盾的结果。

其他重要发现可能包括额外的支持性证据或其他方法的测定，以及可能对论文有意义的额外的结果。并非实验和观察所获得的每个结果都必须在结果部分报道。集中报道最相关的结果，但在选取所报道的结果时，要滤除不相关的结果和压缩有矛盾的结果之间的差异，不要忽略后者。

如果在写作中发现需要收集更多的数据，就应继续数据收集工作。更重要的是做一项彻底的工作，而不是很快地提交一份不完整的稿件。

2) 解释数据

结果指南 4：向读者解释数据

这部分不应只是展示数据，而且要通过向读者展示作为结果的数据时，总结和解释这些数据的含义。只有那些被解释的数据对于读者才有意义。

要清楚地向读者展示成果，并且需要区分数据和结果之间的差别。数据是通过科学实验获取的数值(浓度、吸光度、平均值、百分比的增加)。结果部分要解释数据(如 "Absorbance increased when samples were incubated at 25℃ instead of 15℃")。虽然大多数的数据应该以图表形式表达，但主要结果还应该在正文文字及对所有数据的解释时指出。在阐述解释/效果时，要通过在括号中给出图表序号的形式参考图表中的数据。

例 1：展示数据时缺少解释

Heart rate was 100 beats per minute after digitalis was added (Fig.3).

除非读者是医生，否则他们可能无法把"每分钟 100 次心跳(100 beats per minute)"和"毛地黄(digitalis)"形成任何关联，特别是在没有给出比较值的情况下。需要让读者知道这个值是否高于或低于正常值。

改后的例 1-1：

a. Heart rate increased to 100 beats per minute after digitalis was added (Fig.3).

在修改的例子中，数据得到解释并以结果的形式提出，从而使修改的例子对于读者而言更有意义。

为了使结果对于领域外的非同行也能理解，就需要也给出比较值。

改后的例 1-2：

b. Heart rate increased from 60 to 100 beats per minute after digitalis was added (Fig.3).

当变动的幅度给予一个比较值时("…from 60 to 100…")，这些数据的解释就能够被大多数科学家能理解了。

下面是另一个只给出数据，没有解释或说明的例子。

例 2：展示数据时缺少解释

The sequences for the proteins K 309 and K 415 were compared (Fig.4).

这个例子没有提供对数据的解释。作者既没有解释，也没有分析数据，只仅仅是让读者参照插图。因此，读者不知道这些数据是相似还是不同，因此还需要读者本人来解读数据。所以作者应该给出明确的解释，以免读者得自己找解释。

改后的例 2-1：

When the sequences for the proteins K 309 to K 415 were compared, their C – terminal sections were found to be 90% homologous (Fig.4).

3) 提供数据和统计信息

结果指南 5：将统计信息与数据放在一起，不要使用统计信息代替结果

理论研究论文中的统计信息可能会产生问题。读者常常责怪自己不能理解论文的内容。然而，这种理解问题的真正原因在于统计信息的错误表达。

为避免混淆读者，应对指出的内容给予参照。

例 3：缺少参照的数据

There is a 20% chance of a big earthquake in California.

上述的句子中因为没有给出参照，因而引起读者的许多误解。读者有多种方式解读这句话：加州的 20%地区有发生大地震的可能(20% of the area of California has a big chance for an earthquake)，或在加州发生的地震有 20%为大地震(20% of the earthquake in California are big)，或 20%的可能发生地震的时间是大的(20% of the time the chances for an earthquake are big)。

在给出事件的可能性之前，给出诸如时间和地点的特定参照，可减少混乱。

读者通常获益于文章使用数字或频率来表示统计信息，而不是用可能性或敏感性来表示。许多学生和写作新手列出了烦琐的统计检验结果，而不是在结果部分对实验观测做出描述和说明。在报道统计信息时，应包括描述性统计，如均值、标准差、置信区间、P 值、样本大小，以及诸如卡方或 t 检验等双变量分析，或诸如回归分析的多元分析。要确保向读者解释描述性统计，而不只是在结果部分列出它们。统计分析应为强化数据服务，而不应取代对它们的解释。因此，最好把统计信息放在图的说明或表格中，或在数据描述之后的括号中。

3. 组织结构

1) 总体组织

结果指南 6：结果部分的开头要回答论文的问题

结果指南 7：结果部分的组织按时间顺序或重要性顺序

结果指南 8：强调并标示自己的结果，将文献来源的信息置次要位置

在结果部分的第一段以介绍自己的主要发现作为开始。自己的主要发现指用于论文的总体回答/结论的结果。也可以在第一段落先简要概述主要观察，然后转到主要发现作为开始。在后一种情况下，任何概述都不要超过几个句子，并确保主要发现仍然在第一段出现，因为这是一个重要的位置。

在随后的段落中展示具体的观察。结果的其余部分的总体结构通常或者以时间为顺序，或者以重要性为顺序。使用主题句概述每一个实验。以解释实验目的、简介背景或介绍实验结果作为每一个次级章节的开始。还要考虑用一两段文字来描述实验观察的具体细节。如果具体细节突出或作为例子阐述，会使得读者更好地理解论文。

在整个结果部分，要强调数据及其含义。将对照控制的结果和方法置于重要

位置。

请注意，通常情况下最好在括号中而不是正文文字中提及某个图表，因为图表是支持性的证据而不是结果。

2) 结果中各部分的组织

结果指南 9：将结果分为几个不同的部分

在每个部分指出：实验的目的或背景、实验方法、结果，以及结果的解释（描述性论文可选）。

在组织结果部分时，可考虑将其分为不同的部分。每一部分涉及一组实验。许多部分（不一定是大多数）的篇幅只有一段，有的可能更长。甚至可以考虑将结果分为多个次级小节，每个小节再分为不同的部分。

结果中各部分或段落的信息要很好地组织。这也包括第一个段落。描述特定实验的每个段落应该包含以上四个必要的组成部分。

结果的各部分或段落以主题句开始。主题句通常要说明实验的目的，也可以提供背景信息。紧接目的的是有关实验方法的简短阐述（约半个句子）。目的的撰写形式可能是过渡性短语或从句。紧接实验方法的是实验结果。先介绍重要的、一般性的结果，其后是次要的结果。最后，对结果予以解释，使读者了解其意义。注意给所有的部分予以标示信号。

重要的是，不要只是简单地罗列数据，要向读者解释数据。在结果部分这种解释应只限于 1～2 个句子。避免冗长的解释、推测或结论。将详细的讨论放在讨论部分。

4. 有关结果部分的重要写作原则

1) 单词的选择

结果指南 10：注意单词的选择

应仔细选择结果部分的单词。选择最精确且具有描述性的措辞来反映所要表达的思想，但要保持简单的措辞。除了使用简单、精确的语言，避免行话和冗词外，还应该特别注意在结果部分对以下特定单词和短语的使用。这些词经常被作者（尤其是非英语母语的作者）不经意地使用，但因其暗含不同应加以区别。

did not

慎重选词。在报道结果时使用诸如"did not"的中性描述性词汇，而不是"could not"或"failed to"。例如：We did not detect any insulin production.（中性——没有期望的含义）

clearly/it is clear/obvious

在结果中不要使用"clearly"及类似的具有主观含义的短语。"clearly"使作者似乎显得傲慢，并且看起来好像作者企图影响读者。

significant

在科学中"significant"指"statistically significant"。在写作中如果出现"Flow rate decreased significantly."读者会期望后面的信息为统计方面的细节。如果是报道统计重要性的结果,就要指出显著性水平。

如果不打算提供统计方面的细节,就使用"markedly"或"substantially"代替"significantly"。然而,最好将这些词留到讨论部分。此外,还应记得使用准确数值或参考性数据来量化这些具有定量含义的单词"Flow rate decreased substantially(23%)."

2) 时态

结果指南 11:在结果中使用过去时,但在描述性论文中使用现在时

结果的报道通常使用过去时,因为这些事件和观察是在过去发生的。

例:时态的使用

Imidazole inhibited the increase in arterial pressure.

描述性研究的结果表达例外,因为所描述的内容仍然真实有效,因而使用现在时。其他的例外是一般有效的陈述,即现在仍然是正确的陈述内容或被认为是一般性的规则,仍应该使用现在时。

例:时态的使用

Our results suggest that learning the association between nectar reward and flower type is primarily olfactory mediated.

3) 提示读者的信号

为了强调结果部分的不同单元,可以考虑使用如下表所列的单词信号

目的/问题	实验方法	结果	结果的解释
To determine…	…we did…	We found…	…, indicating that…
To establish if…	X was subjected to…	We observed…	…, consistent with…
Z was tested…	…by/using…	We detected…	…, which indicates that…
For the purpose of…	ABC was performed…		This observation indicates…

最重要的结果应该特别加以强调,以免读者错过。

例:突显重要的结果

Most interestingly, almost half of the newlywed couples (45.5%) start out sharing everyday household tasks equally or with husbands doing even a greater share than their wives.

5. 结果部分的常见问题

结果部分最常见的问题包括以下方面:

- 缺失成分(实验目的、实验方法、结果或相关的解释)
- 过多的实验细节
- 包含了比较、推测和超出结果说明的结论

1) 缺少成分

在结果部分应包括的四个成分(实验目的、实验方法、结果、相关的解释)中，写作新手往往忘记实验目的和结果的解释。如果结果部分缺少这些成分或其他成分，对于读者来说研究发现就不清楚。

2) 不相关的或次要的信息

    结果指南12：省略次要的信息或不相关的一般性内容

不要使用不相关的或次要的信息来混淆读者或作者本人。

例：不相关和次要的信息

It took 2 hr to process 22,000 molecules and 32 hr to screen the entire ChemBridge database.

读者没有兴趣阅读你的工作花费了多长时间。这样的陈述应该删除。

此外，应删除不相关的研究目的的一般性陈述或诸如下列概述性句子。

例：不相关的概述性句子

To present our results, we first list all components of the macromolecule together with their optima and then describe the outcome of their individual omission.

概述句子只会增加混乱。写作具有一致性和良好的组织结构即可，无须解释是如何开展写作的。

除了概述性句子，也应该避免重复图表中所有的数字。相反，描述研究结果并加以括号的形式告诉读者参照相应的图表。

例：告诉读者参照图表

A total of 34 stilbenes were identified in this study, and they are listed in Table 3.

改后的例：

A total of 34 stilbenes were indentified in this study (Table 3).

主要结果应该在文中用文字描述，即使它们已经在图表中展示。

3) 过度描述的实验细节

    结果指南13：避免实验细节

不要在结果部分再次详细地描述实验方法，不要引入在材料与方法部分中没有提到的实验步骤。

4) 包含了比较、推测和超出结果说明的结论

    结果指南14：避免一般性的结论、推测或与他人研究的比较

不要将数据与其他的研究比较，推测可能的机制，或得出一般性的结论。将这些比较、推测和结论放到讨论部分，但是要记得在每一个结果部分后面简要地

加以解释，为讨论做准备。

6. 结果的修改

完成结果部分后(或者应同事的要求帮助编辑修改结果部分时)，可以按下列清单来系统地"拆分"该部分：

(1) 是否报道了所有的主要发现和其他重要发现？
(2) 最重要的结果和说明是否在结果这节开始的部分？
(3) 最重要结果的数据是否在文字中也提及？
(4) 段落内部的组织是否按重要性顺序？
(5) 每个结果或段落包含了所有的成分(实验目的、实验方法、结果和说明)？
  a. 每一个实验的目的是否明确？
  b. 是否提出了实验方法？
  c. 是否有结果的说明？
(6) 是否给出了所有成分(实验目的、实验方法、结果和说明)的提示性信号？
(7) 是否强调了结果？
(8) 是否与数据一起给出了统计信息？
(9) 是否给读者指出了图表？
(10) 是否包括了控制对照的结果？
(11) 是否避免了不相关的和次要的结果？
(12) 是否排除了一般性的结论、推测或与其他研究的比较？
(13) 文献的引用是否正确和必要？

Reference: Angelika H. Hofmann. 科技写作与交流-期刊论文、基金申请书及会议讲演[M]. 任胜利，莫京，安瑞，等，译. 北京：科学出版社，2012.

Unit 11　Environmental Health

Part A　Intensive Reading

Environmental Health

Many aspects of human well-being are influenced by the environment, and many diseases can be initiated, promoted, sustained, or stimulated by environmental factors. For this reason, the interactions of people with their environment are an important component of public health.

In its broadest sense, environmental health is the segment of public health that is concerned with assessing, understanding, and controlling the impacts of people on their environment and the impacts of the environment on them. Even so, this field is defined more by the problems it faces than by the approaches it uses. These problems include the treatment and disposal of liquid and airborne wastes, the elimination or reduction of stresses in the workplace, the purification of drinking-water supplies, the provision of food supplies that are adequate and safe, and the development and application of measures to protect hospital and medical workers from being infected with diseases such as acquired immune deficiency syndrome (AIDS) and severe acute respiratory syndrome (SARS).

Environmental health professionals also face long-range problems that include the effects of toxic chemicals and radioactive wastes, acidic deposition, depletion of the ozone layer, global warming, resource depletion, and the loss of forests and topsoil. The complexity of these issues requires multidisciplinary approaches. Thus a team that is coping with a major environmental health problem may include scientists, physicians, epidemiologists, engineers, economists, lawyers, mathematicians, and managers. Input from experts in these and related areas is essential for the development, application, and success of the control strategies necessary to encompass the full range of people's lifestyles and their environment. Just as the field of public health involves more than disease (for example, health-care management, maternal and child health, epidemiology), the field of environmental health encompasses the effects of the environment on animals other than humans, as well as on trees and vegetation and on natural and historic landmarks. While many aspects of public health deal with the "here and now", many of the topics addressed within the subspecialty of environmental health are concerned with the previously cited impacts of a long-range nature.

Defining the Environment

To accomplish their goals effectively, environmental health professionals must keep in mind that there are many ways to define the environment. Some of the more prominent of these are described here.

The inner versus outer environment

From the standpoint of the human body, there are two environments: the one within the body and the one outside it. Separating them are three principal protective barriers: the skin, which protects the body from contaminants outside the body; the gastrointestinal (GI) tract, which protects the inner body from contaminants that have been ingested; and the membranes within the lungs, which protect the inner body from contaminants that have been inhaled.

Although they may provide protection, each of these barriers is vulnerable under certain conditions. Contaminants can penetrate to the inner body through the skin by dissolving the layer of wax generated by the sebaceous glands. The GI tract, which has by far the largest surface area of any of the three barriers, is particularly vulnerable to compounds that are soluble and can be readily absorbed and taken into the body cells. Fortunately, the body has mechanisms that can protect the GI tract: unwanted material can be vomited via the mouth or rapidly excreted through the bowels (as in the case of diarrhea). Airborne materials in the respirable size range may be deposited in the lungs and, if they are soluble, may be absorbed. Mechanisms for protecting the lungs range from simple coughing to cleansing by macrophages that engulf and promote the removal of foreign materials. Unless an environmental contaminant penetrates one of the three barriers, it will not gain access to the inner body, and even if a contaminant is successful in gaining access, the body still has mechanisms for controlling and/or removing it. For example, materials that enter the circulatory system can be detoxified in the liver or excreted through the kidneys.

Although an average adult ingests about 1.5 kilograms of food and 2 kilograms of water every day, he or she breathes roughly 20 cubic meters of air per day. This amount of air weighs more than 24 kilograms. Because people usually cannot be selective about what air is available, the lungs are the most important pathway for the intake of environmental contaminants into the body. The lungs are also by far the most fragile and susceptible of the three principal barriers.

The personal versus ambient environment

In another definition, people's "personal" environment, the one over which they have control, is contrasted with the working or ambient (outdoor) environment, over which they may have essentially no control. Although people commonly think of the working or outdoor environment as posing the higher threat, environmental health experts estimate that the personal environment, influenced by hygiene, diet, sexual practices, exercise, use of tobacco, drugs, and alcohol, and frequency of medical checkups, often has much more, if not a dominating, influence on human well-being. As may be noted, the personal environment and the lifestyles followed by individuals account for about 70 percent or more of such deaths.

The gaseous, liquid, and solid environments

The environment can also be considered as existing in one of three form-gaseous, liquid, or solid. Each of these is subject to pollution, and people interact with all of them. Particulates and gases are often released into the atmosphere, sewage and liquid wastes are discharged into water, and solid wastes, particularly plastics and toxic chemicals, are disposed of on land.

The chemical, biological, physical, and socioeconomic environments

Another perspective considers the environment in terms of the four avenues or mechanisms by which various factors affect people's health.

1. Chemical constituents and contaminants include toxic wastes and pesticides in the general environment, chemicals used in the home and in industrial operations, and preservatives used in foods.

2. Biological contaminants include various disease organisms that may be present in food and water, those that can be transmitted by insects and animals, and those that can be transmitted by person-to-person contact.

3. Physical factors that influence health and well-being range from injuries and deaths caused by accidents to excessive noise, heat, and cold and to the harmful effects of ionizing and nonionizing radiation.

4. Socioeconomic factors, though perhaps more difficult to measure and evaluate, significantly affect people's lives and health. Statistics demonstrate compelling relationships between morbidity and mortality and socioeconomic status. People who live in economically depressed neighborhoods are less healthy than those who live in more affluent areas.

Clearly, illness and well-being are the products of community, as well as of chemical, biological, and physical, forces. Factors that contribute to the differences range from the unavailability of jobs, inadequate nutrition, and lack of medical care to stressful social conditions, such as substandard housing and high crime rates. The contributing factors, however, extend far beyond socioeconomics. Studies have shown that people without political power, especially disadvantaged groups who live in lower-income neighborhoods, often bear a disproportionate share of the risks of environmental pollution. One common example is increased air and water pollution due to nearby industrial and toxic waste facilities. Disadvantaged groups also suffer more frequent exposure to lead paint in their homes and to pesticides and industrial chemicals in their work.

None of the preceding definitions of the environment is without its deficiencies, and the list is by no means complete. Classification in terms of inner and outer environments or in terms of gaseous, liquid, and solid environments, for example, fails to take into account the significant socioeconomic factors cited earlier or physical factors such as noise and ionizing and nonionizing radiation. As a result, consideration of the full range of existing environments is essential for understanding the complexities involved and controlling the associated problems.

Source: Moeller D W. Environmental Health[M]. Cambridge:Harvard University Press, 2005.

Words and Phrases

epidemiologist ['epɪˌdiːmiˈɒlədʒɪst] n. 流行病学家

subspecialty ['sʌb'speʃəltɪ] n. 附属专业

contaminant [kənˈtæmɪnənt] n. 致污物；污染物

vulnerable ['vʌlnərəbl] adj. 脆弱的

penetrate ['penɪtreɪt] v. 渗透；进入；穿过；看透

sebaceous [səˈbeɪʃəs] adj. 分泌脂质的；皮脂腺的

vomit ['vɒmɪt] v. 呕吐；吐出；涌出；喷出；n. 呕吐；呕吐物

excrete [ɪkˈskriːt] v. 排泄

respirable ['resp(ə)rəb(ə)l] adj. 可呼吸的；能呼吸的

macrophage ['mækrəfeɪdʒ] n. 巨噬细胞

kidney ['kɪdni] n. 肾；肾脏；（食用的）动物腰子

tobacco [təˈbækəʊ] n. 烟草；烟叶

preservative [prɪˈzɜːvətɪv] n. 防腐剂；保护剂；adj.（能）保存的；储藏的；防腐的

ionize ['aɪənaɪz] v.（使）电离

morbidity [mɔːˈbɪdəti] n. 发病率；发病；病态；不健全

mortality [mɔːˈtæləti] n. 死亡数，死亡率；必死性，必死的命运

substandard [sʌbˈstændəd] adj. 不达标的；不合格的

disproportionate [ˌdɪsprəˈpɔːʃənət]

adj. 不成比例的；不相称的；太大（或太小）的

acquired immune deficiency syndrome （AIDS） 艾滋病；获得性免疫缺陷综合征

severe acute respiratory syndrome （SARS） 严重急性呼吸系统综合征

Questions

1. What are the problems that we are facing in environmental health aspect?
2. What are the barriers to separate the inner and outer environment?
3. How can the environment factors affect human health?
4. Why the urban environment becomes more and more important?

Sentence-making

1. contamination, stimulate, epidemiology
2. preservative, member, toxicity
3. morbidity, mortality, maximum
4. respirable, $PM_{2.5}$, cancer
5. socioeconomic, physical, chemical

Part B Extensive Reading

Transforming Environmental Health Protection

In 2005, the U.S. Environmental Protection Agency (EPA), with support from the U.S. National Toxicology Program (NTP), funded a project at the National Research Council (NRC) to develop a long-range vision for toxicity testing and a strategic plan for implementing that vision. Both agencies wanted future toxicity testing and assessment paradigms to meet evolving regulatory needs. Challenges include the large numbers of substances that need to be tested and how to incorporate recent advances in molecular toxicology, computational sciences, and information technology; to rely increasingly on human as opposed to animal data; and to offer increased efficiency in design and costs. In response, the NRC Committee on Toxicity Testing and Assessment of Environmental Agents produced two reports that reviewed current toxicity testing, identified key issues, and developed a vision and implementation strategy to create a major shift in the assessment of chemical hazard and risk. Although the NRC reports have laid out a solid theoretical rationale, comprehensive and rigorously gathered data (and comparisons with historical animal data) will determine whether the hypothesized improvements will be realized in practice. For this purpose, NTP, EPA, and the National Institutes of Health Chemical Genomics Center (NCGC) (organizations with expertise in experimental toxicology, computational toxicology, and high-throughput technologies, respectively) have established a collaborative research program.

EPA, NCGC, and NTP Joint Activities

In 2004, the NTP released its vision and roadmap for the 21st century, which established initiatives to integrate high-through-put screening (HTS) and other automated screening assays into its testing program. In 2005, the EPA established the National Center for Computational Toxicology (NCCT). Through these initiatives, NTP and EPA, with the NCGC, are promoting the evolution of toxicology from a predominantly observational science at the level of disease-specific models in vivo to a predominantly predictive science focused on broad inclusion of target-specific, mechanism-based, biological observations in vitro (Figure1).

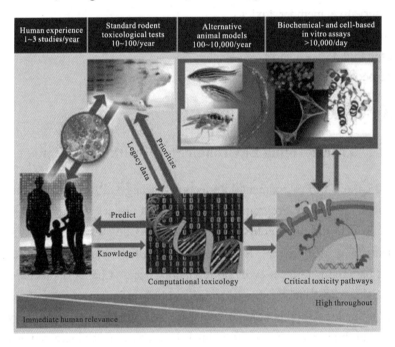

Figure 1 A vision and roadmap for the 21st century on toxicology

Transforming Toxicology

The studies we propose will test whether high-throughput and computational toxicology approaches can yield data predictive of results from animal toxicity studies, will allow prioritization of chemicals for further testing, and can assist in prediction of risk to humans.

Credit: National Institute of Environmental Health Sciences, National Institutes of Health

Toxicity pathways. In vitro and in vivo tools are being used to identify cellular responses after chemical exposure expected to result in adverse health effects. HTS methods are a primary means of discovery for drug development, and screening of >100,000 compounds per day is routine. However, drug-discovery HTS methods traditionally test compounds at one concentration, usually between 2 and 10 μmol, and tolerate high false-negative rates. In contrast, in the EPA, NCGC, and NTP combined effort, all compounds are tested at as many as 15 concentrations, generally ranging from 5 nmol to 100 μmol, to generate a concentration-response curve. This approach is highly reproducible, produces significantly lower false-positive and false-negative rates than the traditional HTS methods, and facilitates multiassay comparisons. Finally, an informatics platform has been built to compare results among HTS screens; this is being expanded to allow comparisons with historical toxicologic NTP and EPA data. HTS data collected by EPA and NTP, as well as by the NCGC and other Molecular Libraries Initiative centers, are being made publicly available through Web-based databases (e.g., PubChem). In addition, efforts are under way to link HTS data to historical toxicological test results, including creating relational databases with controlled ontologies, annotation of the chemical entity, and public availability of information at the chemical and biological level needed to interpret the HTS data. EPA's DSSTox (Distributed Structure Searchable Toxicity) effort is one example of a quality-controlled, structure-searchable database of chemicals that is linked to physicochemical and toxicological data.

At present, more than 2800 NTP and EPA compounds are under study at the NCGC in over 50 biochemical- and cell-based assays. Results from the first study, in which 1408 NTP compounds were tested for their ability to induce cytotoxicity in 13 rodent and human cell types, have been published. Some compounds were cytotoxic across all cell types and species, whereas others were more selective. This work demonstrates that titration-based HTS can produce high-quality in vitro toxicity data on thousands of compounds simultaneously and illustrates the complexities of selecting the most appropriate cell types and assay end points. Additional compounds, end points, and assay variables will need to be evaluated to generate a data set sufficiently robust for predicting a given in vivo toxic response.

In 2007, the EPA launched ToxCast to evaluate HTS assays as tools for prioritizing compounds for traditional toxicity testing. In its first phase, ToxCast is profiling over 300 well-characterized toxicants (primarily pesticides) across more than 400 end points. These end points include biochemical assays of protein function, cell-based transcriptional reporter assays, multicell interaction assays, transcriptomics on primary cell cultures, and developmental assays in zebrafish embryos. Almost all of the compounds being examined in phase 1 of ToxCast have been tested in traditional toxicology tests. ToxRefDB, a relational database being created to house this information, will contain the results of nearly $1 billion worth of toxicity studies in animals.

Another approach to identifying toxicity pathways is exploring the genetic diversity of animal and human responses to known toxicants. The NCGC is evaluating the differential sensitivity of human cell lines obtained from the International HapMap Project to the 2800 compounds provided by NTP and EPA. Similarly, NTP has established a Host Susceptibility Program to investigate the genetic basis for differences in disease response using various mouse strains; cell lines derived from these animals will be evaluated at the NCGC for differential sensitivity to the compounds tested in the HapMap cell lines. The resulting collective data sets will be used to develop bioactivity profiles that are predictive of the phenotypes observed in standard toxicological assays and to identify biological pathways which, when perturbed, lead to toxicities. The ultimate goal is to establish in vitro "signatures" of in vivo rodent and human toxicity. To assist, computational methods are being developed that can simulate in silico the biology of a given organ system.

The liver is the most frequent target of more than 500 orally consumed environmental chemicals, based on an analysis of the distribution of critical effects in rodents by organ, in the EPA Integrated Risk Information System (IRIS). The goal of the recently initiated Virtual Liver project is to develop models for predicting liver injury due to chronic chemical exposure by simulating the dynamics of perturbed molecular pathways, their linkage with adaptive or adverse processes leading to alterations of cell state, and integration of the responses into a physiological tissue model. When completed, the Virtual Liver Web portal and accompanying query tools will provide a framework for incorporation of mechanistic information on hepatic toxicity pathways and for characterizing interactions spatially and across the various cells types that comprise liver tissue.

Targeted testing. The NRC committee report proposes in vitro testing as the principal approach with the support of in vivo assays to fill knowledge gaps, including tests conducted in nonmammalian species or genetically engineered animal models. A goal would be to use genetically engineered in vitro cell systems, microchip-based genomic technologies, and computer-based predictive toxicology models to address uncertainties.

Dose-response and extrapolation models. Dose-response data and extrapolation models encompass both pharmacokinetics (the relation between exposure and internal dose to tissues and organs) and pharmacodynamics (the relation between internal dose and toxic effect). This knowledge can aid in predicting the consequences of exposure at other dose levels and life stages, in other species, or in susceptible individuals. Physiologically based pharmacokinetic (PBPK) models provide a quantitative simulation of the biological processes of absorption, distribution, metabolism, and elimination of a substance in animals or humans. PBPK models are being created to evaluate exposure-response relation for critical target organ effects. These can be combined with models that measure changes in cells in target tissues under different test substance concentrations. This will help to determine the likelihood of adverse effects from "low-dose" exposure, as well as to assess variation among individuals in specific susceptible groups.

Making It Happen

Elements of this ongoing collaboration and coordination include sharing of databases and analytical tools, cataloging critical toxicity pathways for key target organ toxicities, sponsoring workshops to broaden scientific input into the strategy and directions, outreach to the international community, scientific training of end users of the new technologies, and support for activities related to the requirements for national and international regulatory acceptance of the new approaches. It is a research program that, if successful, will eventually lead to new approaches for safety assessment and a marked refinement and reduction of animal use in toxicology. The collective budget for activities across the three agencies that are directly related to this collaboration has not yet been established. Future budgets and the pace of further development will depend on the demonstration of success in our initial efforts. As our research strategy develops, we welcome the participation of other public and private partners.

Source: Collins F S, Gray G M, Bucher J R. Transforming Environmental Health Protection[J]. Science, 2008, 319(5865):906-907.

Words and Phrases

toxicology [ˌtɒksɪ'kɒlədʒi]   *n.* [毒物] 毒物学，[毒物] 毒理学

[例句] At long last countries have begun to accept the results of toxicology tests conducted elsewhere, so these need not be repeated in multiple jurisdictions.

genomics [dʒɪ'nəʊmɪks]   *n.* 基因组学；基因体学

[例句] The president praised scientists working with the Human Genome Project and Celera Genomics Corporation.

vitro ['viːtrəʊ]   *n.* 在试管内

[例句] Used in conjunction with in vitro fertilization, this would eliminate the need for female XO/XX chimeras, although a surrogate mother would still be needed to carry the two-father pregnancy to term.

cytotoxic [saɪtə'tɒksɪk]   *adj.* 细胞毒素的

[例句] Because tumor response to cytotoxic chemotherapy has been poor, chemotherapy is used only for palliation.

rodent ['rəʊdnt]   *n.* [脊椎] 啮齿动物

[例句] Like Plasmodium, which cycles between mosquitoes and man, Toxoplasma cycles between its rodent and feline hosts, living out different phases of its existence in each.

transcriptomics [trɑnskrɪp'təʊmɪks]   *n.* 转录组学

[例句] In the post-genomic era, genomics, transcriptomics, proteomics and metabolomics provide great opportunity for the development of metabolic engineering.

bioactivity [biːəʊ'væktɪvɪtɪ]   *n.* 生物活性；生物活度

[例句] The results of research on traditional medicine of China show the primary bioactivity of nitric oxide blood vessels and the ability to change nitrate to nitrite or nitric oxide.

hepatic [hɪ'pætɪk]   *adj.* 肝的；肝脏色的；治肝病的

[例句] Acute liver failure can cause many complications, including excessive bleeding and increasing pressure in the brain. Another term for acute liver failure is fulminant hepatic failure.

mammalian [mæ'meɪlɪən]   *adj.* 哺乳类动物的

[例句] Indeed, he takes this evolutionary process seriously: he is fond of pointing out to his audiences that they have both "mammalian" and "reptilian" brains that can be employed at different moments.

pharmacodynamics [ˌfɑːməkəʊdaɪ'næmɪks]   *n.* 药效学

[例句] Some drugs act on specific targets in cancer cells only or, like aromatase inhibitors, have different pharmacokinetics/pharmacodynamics in cancer patients compared to noncancer patients.

metabolism [mə'tæbəlɪzəm] *n.* [生理] 新陈代谢

[例句] This is a trend in Hollywood, where many believe coffee not only speeds up their metabolism but also keeps them from eating.

Environmental Protection Agency (**EPA**) （美国）环境保护部
National Toxicology Program (**NTP**) 国家毒理学规划处
National Research Council (**NRC**) 国家研究理事会
toxicity test 毒性试验
molecular toxicology 分子毒理学
chemical hazard 化学危险品
computational toxicology 计算毒理学
high-through-put screening (**HTS**) 高通量筛选
National Center for Computational Toxicology (**NCCT**) 国家计算毒理学中心
disease-specific model 特定疾病模型
transforming toxicology 转化毒理学
prioritization of chemicals 化学物质的优先级
cellular response 细胞反应
adverse health effect 对健康不利的影响
false-negative 假阴性
concentration-response curve 量效曲线
false-positive 假阳性
controlled ontology 控制本体
chemical entity 化学实体
biochemical and cell-based assay 生物化学和细胞化验
titration-based 基于滴定法
end point 端点
assay variable 分析变量
vivo toxic response 体内毒性反应
prioritizing compound 优先化合物
biochemical assay 生化检测
protein function 蛋白质功能
multicell interaction assay 多单元的相互作用分析
primary cell culture 原代细胞培养
toxicity pathway 毒性通路
genetic diversity 遗传多样性
cell line 细胞系
the International Hapmap Project 国际人类基因组单体型图计划
simulate in silico 在计算机中模拟
organ system 器官系统
chronic chemical exposure 慢性接触化学物质
molecular pathway 分子途径
physiological tissue model 生理组织模型
liver tissue 肝组织
genetically engineered 基因工程处理过的
dose-response 剂量反应
extrapolation model 外推模型
internal dose 内剂量
toxic effect 毒性作用
dose level 剂量水平

exposure-response 接触反应
target organ 靶器官
target tissue 靶组织

Questions

1. Why NPA and NTP support a project under the National Research Council (NRC)?
2. What screening method will be used for the test program established by the EPA, NCGC and NTP, and what is the focus of the program?
3. What are the ways to identify toxic pathways?
4. What is the role of dose response and extrapolation model?

Part C Dialogue

Dialogue 1

A: Have you heard the news released today? It's a statement proposed by more than 200 scientists and health professionals.

B: I'm afraid I have never heard of it. What is that?

A: It is said that two ingredients used in thousands of daily necessities to kill bacteria, fungi and viruses linger in the environment and pose a risk to human health. How awful!

B: Daily necessities? Like soaps or detergents? Then it's worth our vigilance!

A: The scientists say the possible benefits in most uses of triclosan and triclocarban are not worth the risk. As you say, the statement urges "the international community to limit the production and use of triclosan and triclocarban and to question the use of other antimicrobials." They also call for warning labels on any product containing triclosan and triclocarban and for bolstered research of the chemicals' environmental toll.

B: I can't agree more. Such ubiquitous danger must be forever removed. Otherwise, the accumulation of harmful substances will lead to even worse consequences.

A: I fully agree with what you are saying. The statement says evidence that the compounds are accumulating in water, land, wildlife and humans is sufficient to merit action.

B: By the way, what are the specific hazards of these substances?

A: Both chemicals affect animals' hormone systems, causing reproductive and development problems. And there is nascent evidence that the impacts may extend to humans as well—having been linked to reduced growth of fetuses, earlier births, and lower head circumference in boys at birth.

B: So what would the government do as a response?

A: In February the EU banned triclosan in hygiene products. U.S. manufacturers are phasing out triclosan from hand soaps after the Food and Drug Administration banned it last year amid concerns that the compound disrupted the body's hormone systems. The FDA noted in the restriction that antibacterial hand soaps were no more effective than non-antibacterial soap and water at preventing illness. However, despite soap bans, triclosan remains in Colgate Total toothpaste, some cleaning products and cosmetics.

B: Unscrupulous manufacturers must be responsible for this!

Reference: http://www.environmentalhealthnews.org/ehs/news/2017/june/triclosan-warning

Dialogue 2

A: Is the environment pollution a big issue in your country?

B: It's always a troublesome issue in my country. The biggest problem is water conservation. Climate in my country is very dry, so water conservation becomes very essential.

A: What are your methods in conserving water?

B: In our country, water is rationed. We can only use a certain amount each month. It means that we can not use some modern household items, like washing machines. As you know, they use too much water.

A: I see. But as far as I'm concerned, the biggest environment problem in my country is air pollution.

B: The air surrounding us seems polluted much heavier than that in my country.

A: We have reduced air pollutant emission in recent years, but cars are still a major source. Factories have gone "green" because of the stricter environmental protection law.

B: The problem is now on a truly global scale. I don't believe that any single country could shake it.

A: You're totally right. There need to be an international response to this problem. People should hang together and face the problem.

B: There are so many environment problems in the world today. Do you think we can really solve them all or not?

A: I hope that world leaders can get together and figure out an action plan, but I doubt it'll happen before too late.

B: We need to solve the air pollution problem before we destroy the atmosphere. There're lots of clean, modern technologies, but companies in developing countries still consider it too expensive introducing environmental technologies. Developing countries put more emphasis on economic development.

A: Everyone is looking at the issue in the short term, rather than the long term. It's exactly the reason why rain forests destruct. Countries and timber companies only think about wood. They're never thinking about the long-term damage to the forests. We should also remember that the forests are an important natural habitat for thousands of species of animal and plant life.

B: In other parts of the world, especially in Africa, there is a problem with desertification. Climate change and over-farming are causing farmland to turn into desert. It means that people cannot grow enough crops.

A: It also means that people sometimes fight over the farmland that remains. Damaging farmland actually leads to conflict between people. So what do you think we should do to protect our environment?

B: As individuals, we can help reduce pollution by cutting back on the amount of energy we use and buying products with plain packaging.

A: Yes, and governments must do their best to save the environment. Do you know any campaigns aimed at reducing pollution and protecting the environment?

B: In our country, a nationwide campaign against plastic bags has been launched to fight environmental pollution. The campaign prohibits shops from gibing out free plastic shopping bags and encourages consumers to use baskets or reusable cloth bags instead.

A: Have you ever thought about joining an organization committed to protecting the environment? You could get involved with projects to improve the environment.

B: I think I'd like to.

A: I had no idea you were such an environmentalist before!

B: To be honest, in order for the earth to continue to be a habitable place, we're all going to have to become more interested in the environment.

Reference: https://wenku.baidu.com/view/f23548e66f1aff00bed51e61.html

Part D Supplementary Vocabulary

Environmental Health

abacterial [eɪbæk'tɪərɪəl] *adj.* 无菌的
accumulation in body tissue 在人体组织中的积累
aggressive pathogen 侵袭性病原体
allergy ['ælədʒi] *n.* 过敏
antibiotic [ˌæntibaɪ'ɒtɪk] *n.* 抗生素
asthma ['æsmə] *n.* 哮喘
bird flu 禽流感
capillarity [ˌkæpɪ'lærɪti] *n.* 毛细管作用，毛细现象
carcinogen [kɑː'sɪnədʒən] *n.* 致癌物，致癌因素
cardiovascular disease 心血管疾病
carrier ['kæriə(r)] *n.* 带菌者，载体
chronic bronchitis 慢性支气管炎
communicable disease 传染病
diarrhea [ˌdaɪə'rɪə] *n.* 腹泻
environmental health hazard 环境健康危害
environmental monitor 环境监测
environmental toxicology 环境毒理学
epidemic [ˌepɪ'demɪk] *n.* 流行病，流行; *adj.* 流行的，传染性的
exposure dose 接触剂量
health care waste 医疗废物
health-related biotechnology 与健康有关的生物学技术
immunological disease 免疫疾病
lethal dose 致死剂量
life expectancy 预期寿命，预计期限
long-term exposure to pollutant 污染物的长期效应
lung cancer 肺癌
malaria [mə'leərɪə] *n.* 疟疾
morbidity [mɔː'bɪdəti] *n.* 发病率
mutation [mjuː'teɪʃn] *n.* 突变
nausea ['nɔːziə] *n.* 恶心
occupational safety 职业安全
offensive industry 有毒工业，有害工业
outbreak of disease 疾病暴发
pathogenic organism 致病生物体
pharmaceutical [ˌfɑːmə'suːtɪkl] *adj.* 制药的，配药的; *n.* 药物
physical examination 体检
poisoning of catalyst 催化剂中毒
potable water 饮用水
potential pollutant 潜在污染物
radiation sickness 辐射病
radioactive damage 放射性损害
sanitary fixture 卫生装备，卫生设备
sore throat 嗓子疼
sterilization [ˌsterəlaɪ'zeɪʃn] *n.* 消毒，灭菌(作用)，绝育
synergistic effect of toxic substance 有毒物质的协同效应
tap water 自来水
typhoid fever 伤寒症
urban heat island 城市热岛
vector of human disease 人类疾病传播媒介

Part E Supplementary Knowledge

如何写英文学术论文的讨论

1. 概述

讨论通常是最难以界定和撰写的部分。许多论文因为讨论部分糟糕而被编辑拒稿。尽管数据可能真实而有趣，但在讨论部分的解释或表达可能模糊这些数据。因此，好的体例和清晰而有逻辑的表达显得尤为重要。

2. 内容

  讨论指南1：指出并解释关键性的结果，给出研究问题的答案
  讨论指南2：总结和归纳
  讨论指南3：牢记谁是潜在的读者

讨论的主要功能是解释关键性的结果并基于这些结果得出结论，换言之，就是回答引言中提出的问题。讨论还应该解释结论是如何得出的，比较和对比自己的结果与已经存在的相关结果，并指出理论意义和实际应用。在清楚地指出自己的研究促进知识进步的同时还应该通过总结和归纳指出论文的意义。

在讨论中，要解释研究工作的新颖性并阐述结果的重要性。还应该包括那些不支持答案的结果的解释，并讨论其他相关的结果和假设。此外，也可讨论研究方法中可能的错误或限制，解释意外的发现，并指出可能的下一步工作。不要再次重复研究工作中的每个细节；在讨论中重复结果部分是缺乏写作经验的作者的通病。另外一个常见错误是又增加一个引言部分。相反，在讨论中应总结和归纳。

根据潜在的读者对象调整讨论的内容，并使其长度不超过实际的需要。如果读者是专业人员，就将讨论的范围限定在读者感兴趣的领域。如果读者对象比较宽泛，可能需要讨论更广泛的意义，并提供更多的一般性的背景。

3. 组织

  讨论指南4：讨论部分采用金字塔形结构
    第一段：基于关键的支持证据开展解释/回答
    中间段落：比较/对比以往的研究
        研究的局限性
        出乎意料的发现
        假设或模型
    最后一段：总结
        重要性/意义

与漏斗形的引言相反，讨论呈如下的金字塔形状。换句话说，讨论是从特殊到一般。

在第一段中，告诉读者主要结论及其含义。在随后的段落中，解释这些结果如何适用于研究领域中的未知方面。在最后一段，总结和归纳研究结果对本领域、相关领域和/或社会的总体重要性。

1) 第一段

讨论部分以重要发现的解释开始，这个解释要给出引言中提出的问题的答案。然后提出支持答案的相关结果、说明和/或其他数据。不要假定读者已经记住了结果或者会去寻找它们，必须呈现给读者完整的**事件**。

因为主要结果的解释是论文中最重要的陈述，因此应该出现在最显著的位置：讨论的第一段。主要研究结果的解释应与引言中提出的研究问题/目的相匹配，并回答引言中的提问。主要研究结果的解释也应在本节的其他重要位置——最后一段中加以重复。

不要把讨论的开始写成第二个引言、结果的总结或重复的二次信息。在讨论的开始句直接给出基于研究发现的答案。如果觉得这样的开始太突然，可以在提出答案前重复一下研究目的或简要的背景。答案前面的陈述应该不超过一两句话。

如果答案在讨论的第一段，读者肯定不会错过。读者通常不会从前往后地阅读整篇论文。更可能的是：浏览摘要，也许阅读一下引言，然后直接跳到讨论。在讨论部分，读者想马上找到研究问题的答案，那就是，该研究主要结果的解释。读者通常不会花时间来阅读全部的讨论。他们通常会阅读第一段，然后最后一段。因此，主要结果和其解释应放在这两个重要的位置。

对阐述有争议的主题或研究发现的论文来说，讨论的开头会有例外(不是对重要结果的解释)。对于有高度争议的主题，首先提出自己的发现并解释其与他人发现的不同，以强调自己的观点。这种组织结构有逻辑地帮助读者准备阅读即将到来的论证。把有争议的研究发现的解释放到这类论文中讨论部分的最后。

如果在讨论的第一段没有提供对主要发现的解释或给出问题的答案，读者通常会停止阅读并直接跳到结论性的段落，以期找到答案。

如果讨论被划分为多个带独立标题的小节，在讨论的第一段提出结论可能会有问题。对于这种讨论，在按小节讨论各重要发现之前可考虑提供一个对研究进行总体概述的段落。将这个概述性段落视为讨论中最重要的段落，基于主要的发现指出主要的、总体性的解释。

2) 中间的段落
  讨论指南 5：根据科学主题或重要性大小来组织论文
  讨论指南 6：将自己的结果与其他已经发表的结果进行比较和对比
  讨论指南 7：解释所有差异、出乎意料的发现和局限性
  讨论指南 8：尽可能地进行概括

  在讨论部分，重要的不仅是把自己的研究发现置于所在领域的背景中，而且要把其与这项工作的重要意义或其他看似相关或不相关的领域联系起来。

  在指出并论证答案以后，要提及其他重要的结果。告诉读者结果的含义及可信程度。为确保讨论是有组织的，而不是随意的，可着眼于引言中所指出的事件的问题/目的。

  处理次要的结果时要像对待主要结果一样：对其进行总结和归纳，而不是简单地重复发现了什么。不要就每项单个结果进行讨论。相反，要列出并解释一般性趋势和倾向，并加以评估。

  论点的表达与个人的风格有关，这体现在第一段和最后一段之间段落的顺序。形成讨论的中间段落时，将讨论的主题按重要性由大向小依次安排，除非有其他理由才需要调整这个顺序。解释研究中所获得的任何新的发现和见解，但不要涉及结果中没有提到的任何新的数据，另外，不要重复任何在论文的其他部分已经表达过的信息。

  指出研究中的任何局限性或意外发现，并提出基于研究发现的新的假说或模型。如果必要，在讨论中用插图阐释复杂的模型。比较并对比自己的发现与先前已发表的相关成果，但要避免试图讨论领域中每一项已有的研究。关注最相关、最重要的研究。客观地说明任何分歧，并且承认和确认他人的工作。给出支持和反对自己结论的观点。只有公正地阐述才能使读者信服。要知道，大多时候都要小心地提出自己的观点，而非武断。

比较和对比

  除了提出并论证答案外，还需要解释研究发现如何适合该主题的已有知识。可以通过比较和对比自己的发现与他人的结果来做到这点。可以采取列表的方法，即分别列出自己的发现和他人的结果，这样也许能够更好地分析并讨论自己的工作与前人工作的所有相似或不同。

  当提到任何不支持自己的答案和结论的结果时，要尽最大可能来解释这些发现。如果能解释其中的矛盾，那么表达出来就是值得的。如果无法解释这些发现，则应该承认"We cannot explain why…"，"Although the reason fox X is not obvious…"。

意料之外的发现

除了比较和对比前人的研究并描述研究的局限性，在讨论部分也应提及意料之外的结果。要留意意料之外的结果。不要一下子就假设实验失败了，或者犯了实验错误。意外的结果可能是重要的——它们可能会导致新的发现并改变研究的重点。

在描述意外发现的时候，在段落的开头就指出这个发现是意料之外的(或奇怪的)，然后尽可能地给予解释。

3) 最后一段

在讨论结束时，应该撰写一段结论性的总结。读者通常期望在科学论文的总结中看到两项内容：对最重要结果的分析和研究工作的意义。最重要的结果分析通常由主要结果的解释来体现，即答案。此外。答案应该与引言中的问题/目的相匹配，并且是讨论的第一段所出现的答案。总结中不应出现新的证据。相反，通过复述答案(主要结果的解释)来完成"宏图"。

研究工作的意义可通过深远的解释和讨论部分结束处的结论来完成。尝试将具体的结果拓展至更广泛的意义。根据研究工作的可信度，研究意义可变化为实际应用和理论意义。通过在最后一段增加实际应用、提供建议、建议下一步工作、提出理论命题等加强论文的重要性。讨论任何理论意义、可能的应用、建议或推测都要基于研究结果。如果提出推测或意义，要立足于坚实的证据，并确保读者明白这些都是作者提出的推测或意义。

结论是论文的主要信息，因此措辞应该十分谨慎。

4. 讨论中的重要写作原则

1) 语气

写作的语气很重要。"初学者"经常做了非常好的工作，但自卑于自己的知识有限和"专家"出色的工作。然而，当你收集到足够多的数据并将其撰写成论文时，你便也成为了专家。要让你的写作传递信心和权威。显示你有关于这个主题的知识，并且为你的结论负责任。不要害怕选择立场。

非英语母语的作者也有担心，有些非英语母语的作者没有意识到他们的语气太固执己见，语言的使用太强硬。请英语母语者，最好是科学编辑来阅读和修改稿件，因为母语者会更加关注语言的细微差别。

如果能够避免的话，不要在讨论中使用第三人称和被动语态。相反，使用第一人称和主动语态，使讨论更加生动有趣。使用"we"完全没有问题。如果是单一的作者，也可以考虑使用"we"。

2) 时态

在讨论中也许会发现一些动词时态的混用。请记住，当指研究发现和完成的行为时,使用过去时态。然而,对于一般有效的陈述及仍然是正确的信息,使用现在时。此外,要使用现在时表述问题的答案和陈述研究意义。

3) 连续性

讨论中的连续性尤其重要,可以使读者阅读整个部分而不会因部分段落的不明白而受到羁绊。为了确保这种连续性,要使用主题句、转接词和关键术语。每个段落的主题句不仅要指出段落的主题或该段信息,而且要指出该段与上段的联系,从而针对性地回答或阐述问题/目的。需要仔细研读论文的读者会通过阅读主题句以快速地识别次级主题。

4) 一种选择:结果与讨论

结果和讨论有时合并成一个部分。在合并后的结果与讨论章节,讨论紧随结果,读者可以即刻理解结果为什么重要,而无须再翻阅前面的内容。请注意,在合并后的结果与讨论章节,所有的结果都必须与讨论一并给出,而不是仅仅总结论文中最重要的结果。

5) 不要包含不相关的信息

作者时常在讨论部分增添一些不相关的信息,在空白处填写某些内容或增加讨论的长度。这类信息转变为结果的总结、第二次引言、对一些将要采取的实验步骤的讨论、局限性,或个人观点。讨论部分宜短不宜长。更重要的是只讨论相关结果,而不是立即填满空间。不相关的和重复的信息会分散读者的注意力,因为它不能让读者清楚地识别什么是重要的,什么是次要的。

6) 研究的重要性要清楚

常见情况是对结果的重要性没有予以讨论或讨论不充分,从而导致读者不明白论文的意义。在结论的最后指出研究发现的意义,如此,就可以成为本研究的专家了。最后一段是本节另外一个重要段落,不要让研究的重要性以外的内容来浪费这个段落。

尽管应该讨论本文的研究如何与已有的研究结果相符合,但关键还是要整体层面上集中概述和总结本研究的解释、比较和对比。

Reference: Angelika H. Hofmann. 科技写作与交流-期刊论文、基金申请书及会议讲演[M]. 任胜利, 莫京, 安瑞, 等, 译. 北京: 科学出版社, 2012.

Unit 12　Other Environmental Area

Part A　Environmental Monitoring

　　Environmental monitoring programs were initially conducted on a local basis and had two basic objectives: (1) to estimate exposures to people resulting from certain physical stresses (such as noise and radiation) and from toxic materials that are being, or have been, released and are subsequently being ingested or inhaled; (2) to determine whether the resulting exposures complied with the limits prescribed by regulations. Such programs were either "source" related or "person" related. *Source-related* monitoring programs were designed to determine the exposure or dose rates to a specific population group resulting from a defined source or practice. *Person-related* programs were designed to determine the total exposure from all sources to a specific population group. The latter were particularly useful in instances where several sources contributed to the exposures.

　　Although programs of these types continue to be important, it is increasingly recognized that assessing risks solely to human health and focusing on problems only on a local scale are inadequate. The purposes and goals of environmental monitoring programs today have expanded far beyond these earlier objectives (Table1). Significantly, it is now accepted that some of these programs should have an *environment-related* component, and that conditions should be examined on a regional, local, and global basis. That is, programs should be designed to assess the impact of various contaminants on selected segments of the environment, including ecosystems, and to evaluate factors that may have wide-scale, long-range effects.

Table 1　Types and purposes of environmental monitoring programs

Type of Program	Purpose
Research related	To determine transfer of specific pollutants from one environmental medium to another and to assess their chemical and biological transformation as they move within the environment; to determine ecological indicators of pollution; to confirm that the critical population group has been correctly identified and that models being applied are accurate representations of the environment being monitored
Based on administrative and legal requirements	
Compliance related	To determine compliance with applicable regulations
Public information	To provide data and information for purposes of public relations

Designing an Environmental Monitoring Program

One of the first steps in designing an environmental monitoring program is to define its objective (what samples are to be collected and where and when) and how the data are to be analyzed (Table 2). The program must be planned not only so that the relevant questions are asked at the right time, but also so that the data necessary to answer these questions are obtained. As a result of the increasing sophistication of our understanding of the environment and accompanying technological developments, it is mandatory that these questions must be addressed on a continuing basis. In fact, the answers to them will never become final. Other attributes of a successful environmental monitoring program are that (1) it is sufficiently inexpensive to survive unexpected reductions in supporting funds; (2) it is simple and verifiable so that it is not significantly affected by changes in personnel; and (3) it includes measurements that are highly sensitive to changes within the environment.

Table 2 Questions to be answered in implementing an environmental monitoring program

Program stage or component	Question
Purpose	What is the goal or objective of the program?
Method	How can the goal or objective be achieved?
Analysis	How are the data to be handled and evaluated?
Interpretation	What might the data mean?
Fulfillment	When and how will attainment of the goal or objective be determined?

Most environmental monitoring programs have at least five stages: (1) gathering background data, (2) identifying and evaluating the various pathways of exposure, (3) collecting and analyzing samples, (4) establishing temporal relationships, and (5) confirming the validity of the results. Essential in every case is the identification of the potentially critical contaminants that might be released, their pathways through the environment, and the avenues and mechanisms through which they may cause population exposures. Equally essential is the collection of samples that are representative and the application of standard methods in their analyses and quality control.

Background data

Before monitoring begins, background information is needed on other facilities in the area, the distribution and activities of the potentially exposed population, patterns of local land and water use, and the local meteorology and hydrology. These data permit identification of potentially vulnerable groups, important contaminants, and likely environmental pathways whose media can be sampled.

People responsible for the background analysis must take into account the type of installation, the nature and quantities of toxic materials being used, their potential for release, the likely physical and chemical forms of the releases, other sources of the same contaminants in the area, and the nature of the receiving environment. This last item includes natural features (climate, topography, geology, hydrology), artificial features (reservoirs, harbors, dams, lakes), land use (residential, industrial, recreational, dairying, farming of leaf or root crops), and sources of local water supplies (surface or groundwater). Results from a monitoring program conducted before a facility begins operation can be used to confirm these analyses and establish baseline information for subsequent interpretation.

Evaluation of pathways of exposure

Contaminants released from an industrial facility may end up in many sections of the environment, and their quantity and composition will vary with time and the nature and extent of facility operation. As a result, discharged materials can reach the public through many pathways. For example, a secondary lead smelter has the potential to release elemental lead and associated compounds into the atmosphere, whereupon they may become an inhalation hazard(Figure 1). The same facility can also release these contaminants to the soil, either directly or through the air, whereupon they may contaminate groundwater and subsequently be taken up by fish and agricultural products. In a similar manner, the milk from cows and the beef from cattle that graze on pastures adjacent to lead smelters can be expected to have a higher-than-normal lead content. Children who play on contaminated earth near such smelters have shown elevated lead concentrations in their blood. Arsenic emitted by copper smelters follows identical pathways of contamination and human exposure.

Tracing the movement of all contaminants through all potential pathways would be physically and economically impossible. Fortunately, in most cases, for example, a nuclear installation, the primary contributors to population dose consist of no more than half a dozen radionuclides that move through no more than three or four pathways. Once these are identified, along with the habits of the people who live or work in the vicinity, it should be possible to identify a "reasonably maximally exposed" group of individuals (the "critical group") whose activities and location would make them likely to receive the largest exposures. That is not to say, however, that unsuspected pathways may not be important. For years, operators of a major nuclear facility in the United Kingdom disposed of low-level liquid radioactive wastes into the Irish Sea, since theoretical evaluations had shown that this would be acceptable. Later, they discovered to their dismay that certain population groups were consuming larger quantities of radioactive material than had been anticipated. The source was seaweed that they made into flour to make bread.

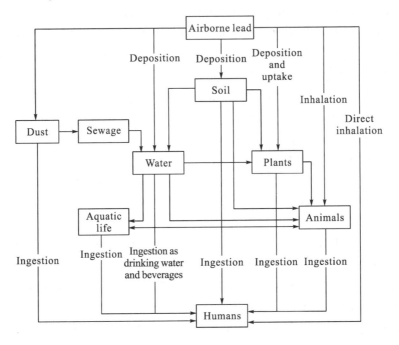

Figure 1　Possible pathways to humans of lead released into the atmosphere

Sample collection and analysis

　　In almost any environmental monitoring programs, trade-offs must be made to obtain adequate coverage of critical contaminant-pathway combinations at satisfactory analytical sensitivities and costs. Samples collected directly from effluent streams that are discharged into the air and water usually contain the largest number of contaminants at the highest concentrations. Analyses of these samples can provide information on the specific contaminants that are being released as well as the amounts anticipated to be present in the neighboring environment. Once sampling and analyses shift to the environment, it is better, under essentially all conditions, to collect and carefully analyze a small number of well-chosen samples taken at key locations to provide a reliable index of environmental conditions than to process larger numbers of poorly selected, non-representative samples. At the same time, there may be a need to measure certain other contaminants because of the history of operations at the site or because of specific concerns of the local population. With the increased interest in assuring that the environment is being adequately protected, it may also be wise to collect and analyze samples that will indicate the range of exposures to plants and animals within various ecosystems.

Because they are faster and less expensive than analyses for specific contaminants, gross measurements of the concentrations of "total suspended particulates" in the atmosphere are sometimes used as a surrogate for, or indicator of, trends in the concentrations of particles in the $PM_{2.5}$ and/or PM_{10} size range. This is acceptable only so long as the relationships between the surrogate and the specific particle size group remain reasonably constant. To ensure that this is the case, the relationship should be verified on an intermittent basis, and especially when the nature of industrial operations or traffic patterns in an area change.

Another factor to be considered is the wide range of impacts resulting from continuing reductions in the permissible limits for various contaminants in essentially every type of environmental media. One of these is the need for more sensitive and more accurate analytical capabilities. Although one way to solve this problem is to collect larger samples, this is not always necessary since technological developments have, in many cases, enabled the more sensitive measurements to be made on existing size samples. At the same time, however, the ability to detect almost any amount of a contaminant, regardless of how small it is, will raise fears in certain segments of the public. Until agreement is reached on what levels of exposure to people are sufficiently small to be considered acceptable, this situation will continue.

Use of standard procedures

Related types of guidance are needed for many aspects of environmental monitoring, including documentation of the adequacy of the cleanup of Superfund sites, those being converted into brownfields, decommissioned commercial nuclear power plants, and facilities that are being remediated by the U.S. Department of Energy. To meet the latter two needs, the EPA and the USNRC, supported by the Departments of Energy and Defense, jointly developed a Multi-Agency Radiation Survey and Site Investigation Manual (EPA and USNRC, 1997; Abelquist, 2003). As in the case of Standard Methods, organizations that apply the techniques described in the cited manual can do so with the knowledge that they are following practices that will be acceptable to regulatory agencies.

Quality assurance requirements

To be effective, an environmental monitoring program must be supported by a sound quality-assurance program. This must include (1) acceptance testing or qualification of laboratory and field sampling and analytic devices; (2) routine calibration of all field-associated sampling equipment and flow-measuring instrumentation; (3) a laboratory cross-check program; (4) replicate sampling on a systematic basis; (5) procedural audits; and (6) documentation of laboratory and field procedures and quality-assurance records.

Source: Moeller D W. Environmental Health[M]. Cambridge:Harvard University Press, 2005.

Words and Phrases

radiation [ˌreɪdɪˈeɪʃn] *n.* 辐射；放射物

mandatory [ˈmændətəri] *adj.* 强制的；托管的；命令的；*n.* 受托者（等于 mandatary）

verifiable [ˈverɪfaɪəbl] *adj.* 能证实的

temporal [ˈtempərəl] *adj.* 时间的；世俗的；暂存的；[语] 表示时间的；*n.* 暂存的事物，世间的事物；世俗的权力；俗事

topography [təˈpɒɡrəfi] *n.* 地貌；地形学；地形测量学

reservoir [ˈrezəvwɑː(r)] *n.* 水库；储藏；大量的储备；储液槽

recreational [ˌrekriˈeɪʃənl] *adj.* 娱乐的，消遣的；休养的

dairying [ˈdeərɪɪŋ] *n.* 制酪业；乳制品业

inhalation [ˌɪnhəˈleɪʃ(ə)n] *n.* 吸入；吸入药剂

pasture [ˈpɑːstʃə(r)] *n.* 草地；牧场；牧草；*vt.* 放牧；吃草

concentration [ˌkɒnsənˈtreɪʃnz] *n.* 关注点；[土壤] 浓聚物（concentration 的复数）

ingestion [ɪnˈdʒestʃən] *n.* 摄取；吸收；咽下

environmental monitoring 环境监测
monitoring program [计] 监控程序
toxic material 有毒物质
source related 源相关
population group 人群组，人口群
assessing risk 风险评估
local scale 局部比例尺
biological transformation 生物转化
ecological indicator 生态指示生物，生态指示种；生态指标
legal requirement 法律要件
meteorology and hydrology 气象和水文
vulnerable group 弱势群体
root crop *n.* [作物] 块根农作物
industrial facility 工业设施；工业企业生产设备
nuclear facility 核设施
cross-checking *n.* 横杆推挡；相互（相互）检验；重复检验；*v.* 反复核对；反复观察（cross-check 的现在分词）

Questions

1. What are the two basic objectives of environmental monitoring program?

2. Who is responsible for the local, regional and national environmental monitoring programs?

3. What is the first step in designing environmental monitoring programs? What are the five stages of environmental monitoring programs?

4. What are the factors that need to be considered for background analysis?

5. Please provide a brief description of the examples used to describe the evaluation of exposure pathways.

Part B   Risk Assessment

In a personal sense, risk can be defined as the probability that an individual will suffer injury, disease, or death under a specific set of circumstances. In the realm of environmental health, this definition must be expanded to include possible effects on other animals and plants, as well as on the environment itself. Knowing that a certain risk exists, however, is not enough. People want to have some idea of how probable it is that they or their environment will suffer and, if they do, what the effects will be. Determination of the answers to these questions involves the science of risk assessment.

Risk assessment ranges from evaluation of the potential effects of toxic and/or radioactive chemical releases known to be occurring to evaluation of the potential effects of releases due to events whose probability of occurrence is uncertain. In the latter case, the risk is a combination of the likelihood that the event will occur and the likely consequences if it does. In essence, the process of risk assessment requires addressing three basic questions:

- What can go wrong?
- How likely is it?
- If it does happen, what are the consequences?

As might be anticipated, there is a spectrum of endpoints that can be considered in evaluating whether a risk is acceptable. Once the endpoint has been defined, and the associated risk has been assessed, it can be expressed in qualitative terms (such as "high", "low", or "trivial") or in quantitative terms, ranging in value from 0 (certainty that harm will not occur) to 1 (certainty that harm will occur). At the same time, it must be recognized that a given assessment provides only a snapshot in time of the estimated risk of a given toxic agent and is constrained by our current understanding of the relevant issues and problems. To be truly instructive and constructive, risk assessment should always be conducted on an iterative basis, being updated as new knowledge and information become available.

Once a risk has been quantified, the next step is to decide whether the estimated risk is sufficiently high to represent a public health concern and, if so, to determine the appropriate means for control. Such control, which falls under the rubric of what is called risk management, may involve measures to prevent the occurrence of an event as well as appropriate remedial actions to protect the public and/or the environment in case the event occurs. Because of the nature of these processes, each step in the risk-management process is accompanied by a multitude of uncertainties. The same is true for risk assessment.

Qualitative Risk Assessment

Although some regulatory agencies seek to develop quantitative risk assessments, the large number of facilities that have the potential for toxic chemical releases make universal application of such assessments impossible. For this reason, the common approach is to apply as an initial step some type of qualitative or semiquantitative assessment. Possibilities include: (1) qualitative characterizations where health risks are identified but not quantified; (2) qualitative risk estimations where the chemicals present are ranked or classified by broad categories of risk; and (3) semiquantitative approaches where effect levels (for example, "no observable effect") are used in combination with uncertainty factors to establish "safe" exposure levels.

Quantitative Risk Assessment

In contrast, the EPA concentrates on quantitative risk assessments. Its goal is to characterize in numerical terms the potential adverse health effects of human exposures to toxic agents. Such assessments involve four primary steps and serve as one of the principal elements of risk assessment and risk management, of which the former is an essential preparatory step to the latter. Fundamental to these two steps is the conduct of the laboratory and field research to provide the necessary input data.

Hazard Identification

Hazard identification is a qualitative determination of whether human exposure to a specific agent has the potential for adverse health effects. This generally requires information on its identity and the outcomes of related mutagenesis and cell-transformation studies, animal research, and human epidemiological studies. Information on the physical and chemical properties of the agent is also required for assessing the degree to which it can become airborne and be inhaled and absorbed into the body and for evaluating its solubility in water and availability for transport through the food chain.

Dose Response Assessment

Dose-response assessment is a quantitative estimate of the hazard potency (power to produce adverse effects) inherent in receiving a dose from a specific toxic agent. If available, dose-response estimates based on human data are preferred. In their absence, information from studies of other animal species that respond like humans may be used. As previously discussed, however, the use of animal data introduces multiple uncertainties into the accompanying risk estimates.

Exposure Assessment

Exposure assessment is an estimate of the extent of exposure to the agent and the accompanying dose to people and the environment. The assessment may be directed to normal releases from a facility, or it may, in the case of accident assessments, involve estimating both the probability of the event and the magnitude of the accompanying releases. Factors considered in performing such assessments include the following:

1. The chemical and physical characteristics of the agent, for reasons similar to those identified in the hazard-identification step. Key parameters include partition coefficients, retardation factors, bioaccumulation factors, and degradation rates. To the extent possible, the values assigned to such parameters should be specific for the system and site being analyzed.

2. Identification and characterization of the person to be protected. This is especially important in cases where it is necessary to protect special groups, such as children and pregnant women, who may be more highly susceptible to a given agent.

3. Recognition of the difference in the exposure measured and the dose that will actually be received by the exposed individuals. Although several people may be exposed to the same agent, the accompanying dose depends on a number of factors. In the case of airborne materials, the dose depends on the age and breathing rate of the person exposed and on whether he/she breathes through the mouth or nose.

Risk Characterization

Risk characterization involves estimating the dose and accompanying adverse risk to people who have been exposed to a specific agent. This process requires integrating the results of the previous processes to produce quantitative estimates of the associated health and environmental risks. Because risk estimates have significant limitations, the EPA requires that in addition to the estimate itself (usually expressed as a number), the risk characterization contains a discussion of the "weight of the evidence" for human carcinogenicity (for instance, the EPA carcinogen classification); a summary of the various sources of uncertainty in the estimate, including those that arise from hazard identification, dose-response evaluation, and exposure assessment; and a report on the range of risks, using the EPA- based risk estimate as the upper limit and zero as the lower limit. Because of the magnitude of the accompanying uncertainties, it is important that those who perform these exercises not be overly conservative in their assumptions. Otherwise, the associated risk estimates may be far in excess of what will be experienced in the real world. For this reason, many public health officials urge that such assessments be designed so that the outcomes are as realistic as possible.

Ecological Risk Assessment

As some of the major initial challenges in risk assessment were resolved, it became apparent that there was a need to address the impacts of pollution on a broader basis, taking into consideration, for example, environmental impacts such as habitat alteration, the loss of biodiversity, and, most important, the overall impacts of the combination of such activities on ecosystems. As a result, the use and application of such assessments have during the 1990s been slowly but steadily expanded. This is illustrated by recognition of the adverse impacts of acid deposition on lakes and forests, the damaging effects of ozone on agricultural crops, and the need to evaluate the long-term potential impacts of global warming on a host of environmental systems.

To support this effort, the EPA has developed a range of guidance documents (EPA, 2002). In response to the guidance provided, the EPA expanded its approach in this field by focusing on the full range of resources at risk, coupled with consideration of the assessment of the impacts of multiple environmental stressors taking place at multiple scales and with multiple endpoints. This led to the publication of the Wildlife Exposure Factors Handbook, which provides data, references, and guidance for assessments of the impacts of toxic chemicals on wildlife species (EPA, 1993), and Guidelines for Ecological Risk Assessment, which is designed to improve the quality and consistency of such assessments among the EPA's program offices and regions (EPA, 1998).

To provide an organizational component for coordinating and directing such activities, the EPA subsequently created the National Center for Environmental Assessment. One of the immediate realizations of the new center's staff was the lack of information on how ecosystems respond to multiple stressors. Until this void could be filled, progress could not be made on the development of techniques for estimating the accompanying consequences. While it was recognized that short-lived stressors may produce transient and frequently immeasurable effects, longer-lived stressors, in contrast, may for a time appear to be without impacts, only to have seemingly minor effects later emerge in the form of damaging endpoints. At this stage, the center's staff is seeking to identify those resources that are most vulnerable and to assign priorities to the research needed to obtain necessary new data (EPA, 2002).

Source: Moeller D W. Environmental Health[M]. Cambridge:Harvard University Press, 2005.

Words and Phrases

anticipated [æn'tɪsə,peɪtɪd]　*adj.* 预期的；期望的；*v.* 预料(anticipate 的过去分词)；盼望
[例句] Fewer have come than we anticipated.

snapshot ['snæpʃɒt]　*n.* 快照；急射；简单印象；*vt.* 给…拍快照；*vi.* 拍快照
[例句] So what can we gather from this snapshot?

instructive [ɪn'strʌktɪv]　*adj.* 有益的；教育性的
[例句] The most instructive findings of the report relate to economic damages from extreme weather events.

multitude ['mʌltɪtjuːd]　*n.* 群众；多数
[例句] There are a multitude of online and offline classes and groups that teach programming.

semiquantitative [semɪ'kwɒntɪtətɪv]　*adj.* 半定量的
[例句] The second screening of the samples was carried out by semiquantitative PCR using gene-specific primers.

category ['kætəgəri]　*n.* 类别；分类
[例句] In this series of articles, I will not discuss each of these capability categories in detail.

mutagenesis [,mjuːtə'dʒenɪsɪs]　*n.* 突变形成；变异发生
[例句] Then microwave mutagenesis study was carried out with the strain.

dose-response [dəus rɪ'spɒns]　*n.* 剂量反应；剂量效应
[例句] Besides, results of most studies suggest a dose-response relation, with greater reductions in type 2 diabetes mellitus risk with higher levels of coffee consumption.

magnitude ['mægnɪtjuːd]　*n.* 大小；量级；[地震] 震级；重要；光度
[例句] Most software projects have requirements backlogs that are too big — often by an order of magnitude.

carcinogenicity [kɑːsɪnəudʒə'nɪsətɪ]　*n.* 致癌力，致癌性
[例句] This indicates the toxicity and potential carcinogenicity of domestic detergents

on fish.
Risk assessment 风险评价
probability of occurrence 发生概率
a spectrum of 连续的，连串的；……的光谱
estimated risk [经] 估计风险
toxic agent 毒剂；有毒媒介；毒性药物
risk management 风险管理
remedial action 补救行动，矫正措施
initial step 起步
broad categories 大类
effect level 有效水平
exposure level 接触程度；暴露水平

numerical term [数] 数值项
hazard identification 危险标志；危害鉴定；灾难识别
qualitative determination [统计] 定性测定
cell transformation [遗] 细胞转化
epidemiological study 流行病学研究
exposure assessment 承受风险的估计
retardation factor 阻滞因数
risk characterization 风险表征
guidance document 指导性文件

Questions

1. What are the basic issues that need to be addressed in the risk assessment process?

2. What kinds of risk assessment terms are mainly discussed in this paper? What is the difference between them?

3. What is the purpose of quantitative risk assessment?

4. What assessment is qualitative and what is the quantitative term in the article?

Part C Remote Sensing

Remote sensing is the act of acquiring information about an object from a distance. Environmental applications of remote sensing typically involve the collection of photographic or electronic images of the earth's surface or atmosphere from airborne or spaceborne platforms. Visual interpretation or computer processing can then be used to analyze these images.

1. Types of Remote Sensing Systems

Several types of remote sensing systems can be differentiated based on the principles employed for measuring electromagnetic radiation. The most common types fall into four broad categories: photographic systems, video graphic systems, multispectral scanners, and imaging radar systems. Within these categories, particular instruments are designed to operate in specific portions of the electromagnetic spectrum. The earth's atmosphere scatters and absorbs many wavelengths of electromagnetic radiation, limiting the portion of the spectrum that can be used for remote sensing.

Photographic systems

Many types of cameras have been used to acquire photographs of the earth's surface from airplanes and from space.

Common formats include 35mm, 70mm, and 9×9inch film sizes, although specialized cameras that employ other film sizes are also used. Film types include black and white panchromatic, black and white infrared, color, and color infrared, covering the visible and near-infrared portions of the electromagnetic spectrum from approximately 0.4 to 0.9 μm. (Photography in the ultraviolet range, from 0.3 to 0.4 μm, is also possible but is rarely done due to atmospheric absorption and the need for quartz lenses.) Once the film has been processed, photographs can be electronically scanned at a variety of resolutions for use in a digital environment. Photographic systems provide relatively high-resolution images, with the nominal scale of a vertical aerial photo-graph being dependent on the focal length of the camera and the flying height of the sensor.

Video graphic systems

Video cameras can be used to record images in analog form on videotape. Video systems have been designed to operate in the visible, near-infrared, and mid-infrared portions of the electromagnetic spectrum. The advantages of video systems include low cost, near-real-time image availability, and the ability to collect and store many image frames in sequence. The primary disadvantage of video is its low spatial resolution, with approximately 240 lines per image for standard video cameras.

Multispectral scanners

MSS systems use electronic detectors to measure electro- magnetic radiation in selected bands of the spectrum from approximately 0.3 to 14 mm, including the visible and near-, mid-, and thermal-infrared regions. These individual bands may be fairly wide (greater than 0.2 mm in width) or quite narrow (less than 0.01 mm in width). The designs used for MSS systems fall into two categories. Across-track scanners employ a rotating or oscillating mirror to scan back and forth across the line of flight. Along-track ("push-broom") scanners use a linear array of charge-coupled devices (CCDs) to scan in parallel along the direction of flight. Distinct subcategories of MSS systems include thermal scanners, which measure emitted radiation in the thermal infrared portion of the spectrum, and imaging spectrometers, or "hyperspectral scanners", which generally collect data in over 100 continuous, narrow spectral bands, producing a complete reflectance spectrum for every pixel in the image. Airborne and satellite MSS systems have become widely used in many environmental science and resource management applications.

Imaging radar systems

Whereas the previous types of remote sensing systems operate in the visible and infrared portions of the electromagnetic spectrum, imaging radar systems operate in the microwave portion of the spectrum, with wavelengths from approximately 1 cm to 1 m. At these wavelengths, radar is unaffected by clouds or haze (shorter wavelength systems are used for meteorological remote sensing). In addition, radar systems are active sensors, transmitting their own radiation rather than passively measuring reflected solar or emitted radiation; thus, they can be operated at any time of day or night. Imaging radar systems are sensitive to the geometric structure and dielectric properties of objects, with the primary determinant of an object's dielectric properties being its liquid water content.

Photographic cameras, video cameras, and multispectral scanners can be operated in a vertical configuration to minimize the geometric distortion of the image, or at an oblique angle to provide a side view of the landscape. Imaging radar systems are not operated vertically, but in a side-looking configuration with a broad range of possible look angles.

2. Environmental Applications of Remote Sensing

Remote sensing has been used for a wide variety of applications in the environmental sciences. Among the earliest uses of remote sensing was geologic mapping, including the discrimination of rock and mineral types, lineament mapping, and identifying landforms and geologic structures. Today, many types of remotely-sensed data are used for geological applications at a variety of spatial scales, ranging from high-resolution aerial photography, to thermal-scanner images, to lower-resolution Landsat images covering large areas.

Agricultural applications of remote sensing are also common. Aerial photography and other remotely-sensed data are widely used as a base for soil mapping, while multispectral and thermal images are used for soil moisture mapping. Imaging radar systems, with their sensitivity to moisture-related dielectric surface properties, can also be used to measure soil moisture. Multispectral visible and infrared data are used for crop classification and assessment, including monitoring the health and productivity of crops, with the goal of predicting yields and identifying areas of crop damage.

In forestry, aerial photographs are used to delineate timber stands and to estimate tree heights, stocking densities, crown diameters, and other variables relating to timber volume. Color infrared photography and multispectral imagery can be used to map forest types and to identify areas of stress due to pest infestations, air pollution, and other causes. Aerial and satellite imagery can be used to map the effects of wildfires, windthrow, and other phenomena in forested regions. Wildlife habitat can be assessed using remote sensing at a variety of scales. High-resolution aerial photography can also be used to assist in wildlife censuses in non-forested areas such as rangeland.

Many aquatic and hydrological applications make use of remote sensing. Water pollution can be monitored using aerial photography or MSS systems, and imaging radar can be used to detect oil slicks. Thermal imagery is used to study currents and circulation patterns in lakes and oceans. Both optical and radar data are used to monitor flooding, including flooding beneath a forest canopy in the case of radar. Wetlands delineation and characterization can both be assisted by remote sensing. Radar systems are used to measure ocean waves, and both radar and optical images have been used to detect sea and lake ice.

Remote sensing is often used to assist in site selection and infrastructure location, urban and regional planning, and civil engineering applications. Aerial photographs are often acquired with a significant overlap between adjacent photos, allowing heights to be measured using the stereoscopic effect. This process is extensively used for topographic mapping and for creation of geometrically-correct orthorectified photographs to serve as base maps for other applications. Radar interferometry is also being used on an experimental basis for topographic mapping.

3. Image interpretation and Analysis

Many environmental applications of remote sensing rely solely on visual image interpretation. In many cases, visual analysis is improved by stereo viewing of overlapping pairs of images. Increasingly, however, some degree of digital image processing is used to enhance and analyze remote sensing data. Simple image enhancement techniques include data stretches, arithmetic operations such as ratioing and differencing, statistical transformations such as principal components analysis, and image convolution, filtering, and edge detection. More complex image processing techniques include automated land use/land cover classification of images using spectral signatures representing different land cover types.

Most remote sensing applications require the collection of some form of reference data or "ground truth", which is then related to features or patterns in the imagery. For example, pixels in a remotely-sensed hyperspectral image might be compared to a series of mineral spectra acquired from ground samples. Ground measurements of soil moisture, crop productivity, or forest leaf-area index (LAI) could be related to observed reflectance in a satellite image using linear regression. Often, ground truth locations are established using the Global Positioning System (GPS) to facilitate the relation to a georeferenced image.

One significant advantage of digital remotely-sensed imagery, whether collected electronically or as scanned photographs, is the ability to use digital data in a geographic information system (GIS). Once a digital image has been georeferenced, it can be combined with a variety of other types of spatial data. This combination of image and non-image data can be used for a wide range of purposes from simple map updates to complex spatial analysis.

Remote sensing is a rapidly changing field, with more than twenty new satellite systems scheduled for launching in the next decade. Major sources of new data will be high-resolution commercial systems and the various sensors comprising the Earth Observing System (EOS).

Source: Pfafflin J R, Ziegler E N. Encyclopedia of Environmental Science and Engineering[M]. Boca Raton:CRC Press, 2006.

Words and Phrases

photographic [ˌfəʊtə'græfɪk]   *adj.* 摄影的；逼真的；（尤指记忆）详细准确的
[例句] I'd been working as a sales rep. for a photographic company.
spaceborne ['speɪsbɔːn]   *adj.* 运载飞船上的；卫星转播的；[航] 航天器上的
[例句] The topography is from digital elevation data collected by the Advanced Spaceborne Thermal Emission and Reflection Radiometer(ASTER) on NASA's Terra satellite.
spectrum ['spektrəm]   *n.* 光谱；频谱；范围；余象
[例句] Birds perceive the world differently to humans, being able to see light in the ultraviolet range of the spectrum.
scatter ['skætə(r)]   *vi.* 分散；散射；使散射；使散开；使散播；*vt.* 分散；散播
[例句] Put all your eggs into one basket and then watch that basket, do not scatter your shot.
panchromatic [ˌpænkrə'mætɪk]   *adj.* 全色的
[例句] In these new designs, half of the total pixels are arranged in a panchromatic checkerboard and used for luminance.

quartz [kwɔ:ts]　　*n.* 石英
[例句] The quartz watch keeps good time.
resolution [ˌrezə'lu:ʃn]　　*n.* [物] 分辨率；决议；解决；决心
[例句] We subscribe to the resolution.
aerial ['eərɪəl]　　*adj.* 空中的，航空的；空气的；空想的；*n.* [电讯] 天线
[例句] After class, he told them about his adventures in aerospace, and he showed them some aerial photographs.
satellite ['sætəlaɪt]　　*n.* [天] 卫星，[航] 人造卫星
[例句] The television programmes from all over the world are bounced off satellites.
meteorological [ˌmi:tiərə'lɒdʒɪkl]　　*adj.* 气象的；气象学的
[例句] It is against the spirit of meteorological forecasting research.
configuration [kənˌfɪɡə'reɪʃn]　　*n.* 配置；结构；外形
[例句] In this article, however, we deal with only one such configuration.
landform ['lændfɔ:m]　　*n.* [地理] 地貌；[地理] 地形
[例句] This small country has an amazing variety of landforms.

remote sensing　遥感
electronic image　电子图像
visual interpretation　[遥感][测] 目视判读；目视判释
electromagnetic radiation　电磁辐射
multispectral scanner　多光谱扫描仪
imaging radar　[雷达] 成像雷达；测绘雷达
electromagnetic spectrum　[电磁] 电磁谱，[电磁] 电磁波谱
high-resolution images　高分辨率图像
focal length　[光][摄] 焦距
near-infrared　近红外(线)
spatial resolution　空间分辨率
thermal-infrared region　热红外区域
oscillating mirror　往复偏转反射镜
charge-coupled devices (**CCDs**)　电荷耦合器件
imaging spectrometer　成像光谱仪
thematic mapper (**TM**)　[测] 专题制图仪
imaging radar system　成像雷达系统
active sensor　[自] 主动式传感器
vertical configuration　垂直地形
aerial photography　*n.* [摄] 空中摄影；空中照相术
remotely sensed data　遥感数据 遥感影像
optical image　[计] 光学图像
stereoscopic effect　体视效应；立体印像
base map　[地质] 底图；基本图；工作草图
image interpretation　[测][遥感] 图像判读，[遥感] 影像判读
visual analysis　目视分析；视觉分析
arithmetic operation　算术运算
edge detection　[测][遥感] 边缘检测
linear regression　[数] 线性回归
global positioning system (**GPS**)

全球位置测定系统
geographic information system (GIS)
地理信息系统

earth observing system (EOS) 地球观测系统

Questions

1. What is remote sensing? What is the remote sensing environment application?
2. What is the principle of classification of remote sensing systems? What are the categories that are usually divided?
3. What areas are remote sensing applied to?
4. What are the significant advantages of digital remote sensing images?

Part D Introduction to the Environmental Impact Assessment Process

Key information

Box 1
The EIA Process
Environmental Impact Assessment (EIA) is the whole process of:
1. gathering environmental information;
2. describing a development or other project;
3. predicting and describing the environmental effects of the project;
4. defining ways of avoiding, cancelling, reducing or compensating for the adverse effects;
5. publicising the project and the Environmental Statement including a clear, non-technical summary of the likely effects, so that the public can play an effective part in the decision making process;
6. consulting specific bodies with responsibilities for the environment;
7. taking all of this information into account before deciding whether to allow the project to proceed;
8. ensuring that the measures prescribed to avoid, cancel, reduce or compensate for environmental effects are implemented.

The "Environmental Statement" (ES) is the report normally produced by, or on behalf of, and at the expense of, the proposer which must be submitted with the application for whatever form of consent or other authorisation is required. It embraces the first four elements of:

1. gathering environmental information;

2. describing the project;

3. predicting and describing the environmental effects of the project;

4. defining ways of avoiding, reducing or compensating for the adverse effects.

It is only one component, albeit a very important one, of the environmental information that must be taken into account by the decision maker.

The "Environmental Information" that must be taken into account by the decision maker includes the Environmental Statement and all the comments and representations made by any organisations or member of the public as a result of the consultations and publicity that must be undertaken in every case. It also includes any further environmental or other information submitted by the proposer which is relevant to the decision.

"The main aim of the EIA Directive is to ensure that the authority granting consent (the "Competent Authority") for a particular project makes its decision in full knowledge of any likely significant effects on the environment. The Directive therefore sets out a procedure that must be followed for certain types of project before they can be given 'development consent'. This procedure known as Environmental Impact Assessment (EIA) is a means of drawing together, in a systematic way, an assessment of a project's likely significant environmental effects. This helps to ensure that the importance of the predicted effects, and the scope for reducing any adverse effects, are properly understood by the public and the competent before it makes its decision."

Key information

Box 2

It is important to appreciate that EIA is not, in itself, a decision making process.

It is a process that is integrated into existing decision making procedures, for example, the consideration of planning applications or applications for an Improvement Order in respect of land drainage, in order to better inform these decisions as to the environmental implications of the project. In this way, it contributes to the wider objectives of sustainable development.

Consequently, an EIA is not usually undertaken in isolation of some other procedures. Indeed, some procedures, such as the control of the intensive use of uncultivated land and semi-natural areas, were only introduced to provide a regulatory process to ensure compliance with the Directive. The main aim is to protect land which has a particular natural heritage or historic environment value or importance. Comments made on EIA cases still need to focus strongly on representations as to whether the project should proceed, or how it should proceed.

> **Key information**
>
> Box 3
> Comments on an Environmentat Statement should be used to support and justify the representations made in respect of whether the project should be given consent, and if so, what conditions or limitations it should be subject to.
>
> Representations may also relate to the quality or compliance of the Environmental Statement.

The advice in Box 3 is fundamental to the process. This Handbook seeks to distinguish between advice on whether the project should be consented and comments on the environmental information to be taken into account. For example, it is perfectly possible that a consultee may find the conclusions of an Environmental Statement to be appropriate and acceptable, but to conclude that the project ought not to be given consent. Equally, a perfectly acceptable project, from a consultee's point of view, could be accompanied by an inadequate Environmental Statement. In the latter case, the consultee would not, of course object to the proposal, but may advise the Competent Authority about the inadequacy of the Environmental Statement.

> **Key information**
>
> Box 4
> EIA should be of benefit to project proposers, decision makers and all of those consulted in the decision making process, including the public. It should help to ensure that development is sustainable, that development does noe exceed the capacity of the environment to accommodate change without long-term harm. It should help to expedite the decision making process and guide the implementation of those projects that do proceed.
>
> Many of the procedures are required by law but the effectiveness of EIA relies substantially on integrity and good practice.

EIA is intended to ensure that the environmental effects of major developments and other projects likely to have significant environmental effects are fully investigated, understood and taken into account before decisions are made on whether the projects should proceed. Fundamental to the process are the statutory requirements for steps 5-8 (Box 1), namely:

1. the publicising of the project and the Environmental Statement, including a clear, non-technical summary of the likely effects, so that the public can play an effective part in the decision making process;

2. consultation with specific bodies with responsibilities for the environment;

3. taking all of this information into account before deciding whether to allow the project to proceed;

4. ensuring that the measures prescribed to avoid, reduce or compensate for environmental effects are implemented.

Key information

Box 5

In practice, the whole EIA process should be an iterative one (repeated until the best solution has been found), with sometimes complex links back to earlier steps and a continuous process of assessment and reassessment until the best environmental fit is achieved and / or environmental effects cannot be reduced further.

The process can be broken down into a series of stages and steps and summarised in Table 1 below. Whilst the 4 main stages will normally follow consecutively, the steps within each stage could be undertaken concurrently or in a different order.

Table 1   Key Stages and Steps in the EIA Process

Stage		Step
Stage 1: Before Submission of the Environmental Statement	C1	Deciding whether EIA is required (screening)
	C2	Requiring submission of an Environmental Statement
	C3	Preliminary contacts and liaison
	C4	Scoping the Environmental Statement
	C5	Provision of Information
	C6	Describing baseline environmental information
	C7	Predicting environmental impacts
	C8	Assessing the significance of impacts
	C9	Mitigation measures and enhancement
	C10	Presenting environmental information in the Environmental Statement
Stage 2: Submission of Environmental Statement and Consideration of Environmental Information	D1	Submission of Environmental Statement and project application
	D2	Consultation and publicity
	D3	Liaison with the Competent Authority and the Proposer
	D4	Wider Consultation and dissemination
	D5	Transboundary environmental effects
	D6	Requesting more information or evidence

Continued

Stage	Step	
	D7	Modifications to the project
	D8	Further Information and Supplementary Environmental Statements
	D9	Reviewing the Environmental Statement
	D10	Formulating a Consultation Response
	D11	Planning Permission in Principle and Approval of Reserved Matters specified in Conditions
Stage 3: Making the Decision	E1	Adopting the precautionary principle
	E2	Relationship of EIA with the Development Plan and other Procedures
	E3	Guaranteeing commitments and compliance
	E4	The Decision of the Competent Authority
Stage 4: Implementation (For each of the pre-construction, construction, operation, decommissioning and restoration stages)	F1	Compliance and implementation
	F2	Monitoring
	F3	Review, reporting, reassessment and remedial measures

Source: A handbook on environmental impact assessment- Guidance for Competent Authorities, Consultees and others involved in the Environmental Impact Assessment Process in Scotland. Scottish Natural Heritage, 2013. www.snh.gov.uk.

Words and Phrases

compensate ['kɒmpenseɪt] *vi.* 补偿；抵消；*vt.* 赔偿；付报酬
consult [kən'sʌlt] *vt.* 查阅；商量；向…请教；*vi.* 请教；商议；当顾问
proposer [prə'pəʊzə(r)] *n.* 要保人；申请人，提案人
directive [də'rektɪv] *n.* 指示；指令；*adj.* 指导的；管理的
uncultivated [ʌn'kʌltɪveɪtɪd] *adj.* 野生的；未经耕作的，无教养的；野蛮的
accommodate [ə'kɒmədeɪt] *vi.* 适应；调解；*vt.* 容纳；使适应；供应；调解
iterative ['ɪtərətɪv] *adj.* 重复的，反复的，迭代的；*n.* 反复体

dissemination [dɪˌsemɪ'neɪʃn] *n.* 宣传；散播；传染（病毒）
liaison [lɪ'eɪzn] *n.* 联络；（语言）连音
environmental impact assessment (EIA) 环境影响评价
environmental statement 环境报告
in isolation 孤立地；绝缘
semi-natural area 半自然区
the competent authority 主管部门
statutory requirement 法定条件
planning permission 建筑许可证
precautionary principle 预防原则；预警原则

Questions

1. What are the main steps of the environmental assessment?
2. What are the main contents of the Environmental Statement?
3. Can the environmental assessment be done by the private?
4. The process of EIA is likely to be infinitely repeated, so what is the end?
5. What does the EIA apply to the aspects of environmental monitoring?

Part E Supplementary Knowledge

Supplementary Knowledge 1

英文学术论文中的参考文献和致谢

大多数情况下，你的科研成果建立在前人的研究基础上。尽管直接引用在科技写作中比较少见，但是对于原始资料的改写却很常见。无论任何情况下，每当你使用了他人的思想和成果，就需要在正文里注明出处，并将其列在文章最后的文献列表里。参考文献不仅是对他人贡献的一种承认，而且能够进一步引领读者获取其他感兴趣的文献中的信息。另外，文献还给编辑提供了文章可能的评审者，并显示了作者对本专业领域的熟悉程度。

选择文献

　　文献指南 1：选择最相关的文献
　　文献指南 2：用原始文件核实你的参考文献

大多数作者会收集超出稿件所需要的文献。通过使用关键词、作者、来源、作者所在单位、被引作者、被引文章和被引年代来搜索 MEDLINE、SCOPUS、BIOSIS 和 Web of Science 这一类在线数据库，可以很容易获取潜在的相关文献。

从所有文献中立刻筛选出相关文献比较困难。评述文章因包含大量信息，所以含有很多文献，而研究论文则不同，它应当只列出最相关的文献，并且为使读者容易阅读，最好将文献数量控制在最小（平均 20~40 个）。

最相关文献包括最重要的和最容易获取的文献。此类文献通常是期刊文章、书籍和博士学位论文。需要注意的是博士学位论文通常不太容易获取。类似地，会议摘要、会议论文集、个人通信和未出版的数据可以在论文中以括号引用，但这些文献通常不列在文后的参考文献里，因为读者不认为它们是相关或有效文献。因此，它们只能用来支持研究结果，而不应用来支持任何重要的结论。

为降低文献数量，引用原始文章并选择最重要的、水平最高的或最近的论文，而不是列出关于某个话题的所有论文。引用第一手资料来证实特定的研究结果，此类资料是经过同行评审出版的科学家的原始数据、结果和理论。在主题的概述中，也可使用二次或三次文献。二次文献引用并建立在第一手资料上，对其进行讨论或概述，例如评述文章。三次文献概括分析第一手资料，意在提供关于某个主题的全面概述，例如教科书。

当你撰写研究论文、评述文章或项目申请书时，考虑在适当的地方引用评述文章。但是要注意评述文章有时引用不正确的文献或有歪曲误解数据的情况。在引用评述文章或三次文献中的信息时，务必核实其原始出处。

核实文献

文中的引用、参考文献列表和你引用的信息必须准确无误。文献部分的错误率通常很高。所以，你需要核实它们的原文件。为避免错误引用文献或信息，务必阅读所有你引用的文献。另外，保证文中引用的所有文献都包含在文后的文献列表里，并且文后列表里的所有文献都在文中引用。同时，保证引用和文献的格式符合作者须知里的要求。如果直接引用已发表论文中的原句，需要使用引号并且不能做任何改动。如果改写原句，要注意你的表述不能歪曲作者的意思。

帮助你整理文献的电脑软件非常有用。这些软件包括但不限于 EndNote 或 Reference Manager。强烈推荐使用这些软件组织、记录和编排文献。从一开始就整理文献能够为你节省大量时间，省去很多麻烦。没有什么比费力地输入、替换或修改文献列表更加恼人的。从图书馆下载文献的那一刻起就开始使用文献整理软件。

文内引用

形式和顺序

    文献指南 3：关于文献引用的细节须遵照期刊的风格
    文献指南 4：用正确的格式和顺序引用文献

文献以两种方式出现在论文中：文内引用和文后的文献列表。文献以简短的形式被引用在文内，例如以姓名和年份。文后的文献列表则给出完整的文献信息。

无论是文内引用还是文后列表，不同的期刊对文献的引用有不同的要求。查看目标期刊的作者须知来了解如何引用文献，遵守作者须知里关于文献引用的所有细节。

文内引用的一般格式有"（作者，年份）""（文献号）"或"文献号"。

例：a. Vit-E, fat-soluble vitamin, requires micellar formation for absorption and is transported in the intact animal via lymphatics (Hollander et al., 1976).

  b. Vit-E, fat-soluble vitamin, requires micellar formation for absorption and is transported in the intact animal via lymphatics (8).

如果同时引用多个文献，按时间顺序在文内和文后列出文献。如果文内引用采用姓名和年份的方式,则以时间顺序在文内列出多个文献,再以字母顺序在文后列出这些文献。如果文内引用采用文献号,则以数字顺序在文内和文后列出文献。

只有一个作者的文献,通常引用此作者的姓名或姓。

例:…described by Popi (18,20).

有两个作者的文献,引用两个作者的姓名或姓。

例:Daniles and Ebert (9) reported XYZ.

有三个或三个以上作者的文献,只引用第一作者的姓名或姓,后面加"等"或"et al."。

例:…has previously been reported (Brown et al., 1999a; Brown et al., 1999b; Liu et al., 2003).

(注意不同的期刊因其风格差异可能有不同的引用规则。例如:如果你的目标期刊采用 APA 格式,"et al."只在文章有 6 个或 6 个以上作者时才在文内使用。因此,一定要仔细查看目标期刊的作者须知和已发表的论文。)

如果在文内引用会议摘要、会议论文集、个人通信或未发表的数据,应当以括号注明出版状态。以下是说明出版状态的一些范例用语:

manuscript submitted　　manuscript in preparation
unpublished data　　　　data not shown
personal communication　manuscript in press

文内引用的位置

文献指南 5:在句中正确的位置放置文献

文献可以各种方式放入文中,但是必须具体有意义。大体来说,根据你想强调的内容,文献可在引用的内容或作者姓名之后出现:科学还是科学家。

例:a. Starfish fertilization is species specific (17).

b. Peterson (17) reported that starfish fertilization is species specific.

不要将文献放置在某项内容的中间或某项研究的概况信息之后,例如:"in a recent study"或"has been reported"之后。

例:In a current study (16), proteins expressed by the Epstein-Barr virus have been classified into three general groups according to their kinetics and synthesis requirements.

改后的例:In a current study, proteins expressed by the Epstein-Barr virus have been classified into three general groups according to their kinetics and synthesis requirements (16).

但是,文献并不一定要放置在句尾,例如:为区分你的研究和他人的研究,将文献放在相应的研究成果后面,而不是整个研究后面。

同时，注意一个句子中不同内容点的文献需要放在适当的位置，而不是将所有文献一同放在句尾。

抄袭

文献指南6：确保自己没有抄袭

在学术著作中不注明信息来源的行为叫抄袭，这是一种学术不端行为。承认所有你所引用的内容是对其他作者的一种道义责任，即使你没有照抄原文。

为避免抄袭，你需要了解哪些行为构成抄袭，它们包括：

- 引用资料不注明出处(这是最明显的抄袭行为)。
- 借用他人的思想、概念、结果和结论，且不注明出处，将之视为己有，即使这些内容大部分被改写。
- 总结和改写他人的著作但不注明出处。

以上规则适用于文字和视图信息。

被公认为常识的事实不需注明出处。常识是到处都可以找到的信息，并可能众所周知。但是，不被周知的信息和用来解释事实的想法需注明出处。

模仿和借用本身不构成抄袭行为。撰写学术作品时引用别人的思想非常合理并且不可避免，但是必须注明出处。

改写

文献指南7：记录思想和文献
文献指南8：学会改写

改写即借用他人的思想，将之用自己的语言表达出来。这可能是在写作中吸收他人的研究成果时用得最多的技能。尽管你用自己的语言进行了改写，但是也必须在文中句末或想法后面注明出处。

区分改写和抄袭非常重要。只替换原文的一两个词或只改变原句结构不是改写而是抄袭。在科技写作中，避免抄袭的最好方法是在收集和使用信息时做到以下几点：

- 记录下每一条你认为有可能在论文中用得上的文献和准备引用的内容。
- 用文献管理软件详细记录文献来源，例如：Endnote 或 Reference Manager。
- 如果摘抄原文，需使用引号，但是直接引用在科技写作中并不多见。当你想引用原文中的部分细节或者不按照原文的顺序引用某些细节，就需要改写。
- 用自己的语言逐项列出最重要的思想。
- 记笔记的时候把书合上，这样迫使你用自己的语言描述书中的思想。
- 写作时重新核对文献和引用的内容是否正确。

与研究论文的其他部分不同，许多文章的材料与方法部分都极其相似，这主要是因为描述这些除了变量不同，技术和体系基本上相同的方法就只有那么几个。

在论文这部分使用类似的短语并把变量替换成自己的不应被看作是抄袭。因此，不用绞尽脑汁地思考如何用新的方式描述同样的步骤。

对于写作类似这样的材料与方法部分的段落，参考收集其他文章此部分的范例用语不失为一个好方法。但并不提倡大篇幅的摘抄，只建议在论文里借用单个的范例短语和表达方式。

科技论文中的文献

文献指南 9：在科技论文中适当的位置放置参考文献

摘要	避免在摘要中引用任何文献，从引言部分开始引用文献
引言	只引用最相关的文献。尽管需要多少背景信息取决于读者，但是不要进行综述。只引用最近、最重要、一流的，尽可能接近第一手的文献。在合适的情况下考虑引用评述文章。
材料与方法	为研究中使用的材料或方法引用第一手文献，包括发表在大范围发行的期刊上的方法，而不是详述那些方法的细节。
结果	需要引用文献的陈述通常不在结果部分，而在讨论部分，例如：与前人研究的对比。但是，如果一个简单的对比不适合在讨论中出现，也可以写在结果汇总，这样就需要引用文献。
讨论	在讨论部分，尽管研究结果是主要话题，但你需要在一个广泛的范围内来讨论结果。这意味着你需要引用文献来对比研究结果，参考其他研究对结果的解释，或借用其他文章来说明结果的重要性。

文献列表

列出文献

文献指南 10：文献列表的细节参照期刊的风格

只有在文中引用过的文献才能列在文后的参考文献列表里。通常情况下，会议摘要、会议论文集、个人通信和未发表数据不是公共资源，所以，即使文中引用了它们，也不在文献列表里。如果某篇论文已被期刊录用但还未印刷出版，可将其列入文献列表，提供作者姓名、标题、期刊名（如有卷号，也标上），最后标明"（in press）"。

大多数期刊的文献都有自己的风格。务必要仔细查看目标期刊的作者须知，遵照其风格编排文献。关于文献编排风格的细节包括：是否列出文章标题，是否列出起止页码，作者姓名首字母的位置（在姓前面或后面），出版年份的位置（在作者姓名后面，期刊名后面，或本条文献最后），以及标点符号的使用。电脑软件可以实现以各种风格编排文献（Endnote 或 Reference Manager）。注意在文献列表中，姓和名的顺序与英文姓名的顺序正好相反：姓、名或名字首字母、中间名首字母。这个习惯与一些外国期刊的习惯不同。如果在文中使用"（作者，年）"方式，文献列表的文献往往以字母顺序排列，并且不编号。如果在文中使用（数字制）方式，文献列表中的文献将以在文中出现的先后顺序编排。

引用网上资源

文献指南 11：了解如何引用和列出互联网上的文献

在引用网上资源之前，你应该查看目标期刊的作者须知和近期出版的刊物。并不是每个期刊都允许引用网上资源，但这是个正在发展的领域，所以需要事先查看目标期刊的政策。作者通常能够决定选择何种风格。但是，要保证你选择的风格不与目标期刊的要求相冲突。

致谢

致谢指南：列出所有对论文有重要帮助但不足以列为作者的人

概述

感谢所有提供基金、材料、经济或技术支持的组织和个人。如果有可能，具体说明支持的类型。同时，感谢提供想法、信息、建议和对论文的写作和编辑有帮助的人。

感谢那些特地帮助你的人，但不用感谢只是在履行职责的人。要在致谢里提到某人的姓名，需要获得此人的许可和其对致谢措辞的认同。这样可以维护友谊，避免尴尬的情况，并保证未来的合作。

致谢部分通常先列出知识上的贡献，然后是技术支持、材料的提供、有帮助的讨论、稿件的修改和准备。最后，列出有基金、拨款、奖学金或其他经济支持。

竞争性利益声明

有些文字致谢部分的最后一句话是关于利益冲突的声明。当作者(或作者所在机构)、评审者或编辑之间有影响他们行为的(偏见)经济或个人关系，就存在利益冲突。这种关系也被称为双重承诺、利益竞争或诚信对抗。这些关系的程度各不相同，小到可以忽略其对判断的潜在影响，大到对判断产生极大的潜在影响，但不是所有关系都代表真正的利益冲突。冲突可能因其他原因产生，例如：学术判定、经济关系、个人关系、学术竞争和求知欲。同行评审和出版过程中的所有参与者都必须公开所有可视为潜在利益冲突的关系。这种关系的公开对编者按和评述文章来说尤其重要，因为在这类文章中发现偏见比在原创性研究论文中更困难。利益冲突声明通常出现在致谢部分或单独出现在其前后。

利益冲突声明包括以下几种可能的措辞方式：

a. The authors declare no competing financial interests.

b. No potential conflict of interest relevant to this article was reported.

c. Dr. A. reports serving as a consultant to C. Clinic, and serving as a consultant for a patent-infringement case involving U.S. patent xx/yyy, zzz, in the treatment of…

Reference: Angelika H. Hofmann. 科技写作与交流-期刊论文、基金申请书及会议讲演[M]. 任胜利, 莫京, 安瑞, 等, 译. 北京: 科学出版社, 2012.

Supplementary Knowledge 2

实验室常用英语

1. 常用实验仪器名称中英文对照表

Amino Acid Analyzer 氨基酸组成分析仪
Atomic Absorption Spectroscopy 原子吸收光谱仪
Atomic Emission Spectrometer 原子发射光谱仪
Atomic Fluorescence Spectroscopy 原子荧光光谱仪
Automatic Analyzer for Microbes 微生物自动分析系统
Automatic Titrator 自动滴定仪
Bechtop 超净工作台
Biochemical Analyzer 生化分析仪
Bio-reactor 生物反应器
Centrifuge 离心机
Chemiluminescence Apparatus 化学发光仪
CO_2 Incubator CO_2 培养箱
Conductivity Meter 电导仪
Constant Temperature Circulator 恒温循环泵
DNA Sequencer DNA 测序仪
DNA Synthesizer DNA 合成仪
Electro Microscopy 电子显微镜
Electrolytic Analyzer 电解质分析仪
Electron Paramagnetic Resonance Spectrometer 电子顺磁共振波谱仪
Electrophoresis System 电泳仪
Electrophoresis 电泳
Energy Disperse Spectroscopy 能谱仪
Fermenter 发酵罐
Flow Analytical and Process Analytical 流动分析与过程分析
Flow Cytometer 流式细胞仪
Fraction Collector 部分收集器
Freeze Drying Equipment 冻干机
FT-IR Spectrometer 傅里叶变换红外光谱仪
FT-Raman Spectrometer(**FTIR-Raman**)傅里叶变换拉曼光谱仪
Gas Analysis 气体分析
Gas Chromatograph 气相色谱仪
GC-MS 气相色谱-质谱联用仪
Gel Permeation Chromatograph 凝胶渗透色谱仪
High Pressure/Performance Liquid Chromatography 高压/效液相色谱仪
Hybridization Oven 分子杂交仪
ICP-MS ICP-质谱联用仪
Inductive Coupled Plasma Emission Spectrometer 电感耦合等离子体发射光谱仪
Instrument for Polymerase Chain Reaction(**PCR**)PCR 仪
Inverted Microscope 倒置显微镜
Ion Chromatograph 离子色谱仪
Isotope X-Ray Fluorescence

Spectrometer 同位素 X 荧光光谱仪
LC-MS 液相色谱-质谱联用仪
Mass Spectrometer 质谱
Nuclear Magnetic Resonance Spectrometer 核磁共振波谱仪
Optical Microscopy 光学显微镜
Particle Size Analyzer 粒度分析仪
PCR Amplifier PCR 扩增仪
Peptide synthesizer 多肽合成仪
pH Meter pH 计
Polarograph 极谱仪
Protein Sequencer 氨基酸测序仪
Scanning Probe Microscopy 扫描探针显微镜
Sensor 传感器
Sequencer and Synthesizer for DNA and Protein DNA 蛋白质的测序和合成仪
Shaker 摇床
Surface Analyzer 表面分析仪
Thermal Physical Property Tester 热物理性能测定仪
Ultrahigh Purity Filter 超滤器
Ultra-low Temperature Freezer 超低温冰箱
Ultrasonic Cell Disruptor 超声破碎仪
Ultraviolet Detector 紫外检测仪
Ultraviolet Lamp 紫外观察灯
UV-Visible Spectrophotometer 紫外-可见光分光光度计
Voltameter 伏安仪
Water Test Kits 水质分析仪
X-Ray Diffractometer X 射线衍射仪
X-Ray Fluorescence Spectrometer X 射线荧光光谱仪

2. 化学分析英文缩写

AAS	原子吸收光谱法	CZEP	毛细管区带电泳法
AES	原子发射光谱法	DDTA	导数差热分析法
AFS	原子荧光光谱法	DIA	注入量焓测定法
ASV	阳极溶出伏安法	DPASV	差示脉冲阳极溶出伏安法
ATR	衰减全反射法	DPCSV	差示脉冲阴极溶出伏安法
AUES	俄歇电子能谱法	DPP	差示脉冲极谱法
CEP	毛细管电泳法	DPSV	差示脉冲溶出伏安法
CGC	毛细管气相色谱法	DPVA	差示脉冲伏安法
CIMS	化学电离质谱法	DSC	差示扫描量热法
CIP	毛细管等速电泳法	DTA	差热分析法
CLC	毛细管液相色谱法	DTG	差热重量分析法
CSFC	毛细管超临界流体色谱法	EAAS	电热或石墨炉原子吸收光谱法
CSFE	毛细管超临界流体萃取法		
CSV	阴极溶出伏安法	EIMS	电子碰撞质谱法

ELISA	酶标记免疫吸附测定法		发射光谱法
EMAP	电子显微放射自显影法	ICP-MS	电感耦合等离子体质谱法
EMIT	酶发大免疫测定法	IDA	同位素稀释分析法
EPMA	电子探针X射线微量分析法	IDMS	同位素稀释质谱法
ESCA	化学分析用电子能谱学法	IEC	离子交换色谱法
ESP	分光光度法	INAA	仪器中子活化分析法
ETA	酶免疫测定法	IPC	离子对色谱法
FAAS	火焰原子吸收光谱法	IR	红外光谱法
FABMS	快速原子轰击质谱法	ISE	离子选择电极法
FAES	火焰原子发射光谱法	ISFET	离子选择场效应晶体管
FDMS	场解析质谱法	LAMMA	激光微探针质谱分析法
FIA	流动注射分析法	LC	液相色谱法
FIMS	场电离质谱法	LC-MS	液相色谱-质谱法
FNAA	快中心活化分析法	MECC	胶束动电毛细管色谱法
FT-IR	傅里叶变换红外光谱法	MEKC	胶束动电色谱法
FT-MS	傅里叶变换质谱法	MIP-AAS	微波感应等离子体原子吸收光谱法
FT-NMR	傅里叶变换核磁共振谱法		
Gamma	射线发射光谱法	MIP-AES	微波感应等离子体原子发射光谱法
GC	气相色谱法		
GC-IR	气相色谱-红外光谱法	MS	质谱法
GC-MS	气相色谱-质谱法	NAA	中子活化法
GD-AAS	辉光放电原子吸收光谱法	NIRS	近红外光谱法
GD-AES	辉光放电原子发射光谱法	NMR	核磁共振波谱法
GD-MS	辉光放电质谱法	PAS	光声光谱法
GFC	凝胶过滤色谱法	PC	纸色谱法
HAAS	氢化物发生原子吸收光谱法	PCE	纸色谱电泳法
HAES	氢化物发生原子发射光谱法	PE	纸电泳法
HPLC	高效液相色谱法	PGC	热解气相色谱法
HPTLC	高效薄层色谱法	PIGE	粒子激发
IBSCA	离子束光谱化学分析法	PIXE	粒子激发X射线发射光谱法
IC	离子色谱法	RHPLC	反相高效液相色谱法
ICP	电感耦合等离子体	RHPTLC	反相液相薄层色谱法
ICP-AAS	电感耦合等离子体原子吸收光谱法	RIA	发射免疫分析法
		RPLC	反相液相色谱法
ICP-AES	电感耦合等离子体原子	SEM	扫描电子显微镜法

SFC	超临界流体色谱法	TGA	热重量分析法
SFE	超临界流体萃取法	TGC	薄层凝胶色谱法
SIMS	次级离子质谱法	TLC	薄层色谱法
SIQMS	次级离子四极质谱法	UPS	紫外光电子光谱法
SP	分光光度法	UVF	紫外荧光光谱法
SP(M)E	固相(微)萃取法	UVS	紫外光谱法
STEM	扫描投射电子显微镜法	XES	X射线发射光谱法
STM	扫描隧道电子显微镜法	XPS	X射线光电子光谱法
SV	溶出伏安法	XRD	X射线衍射光谱法
TEM	投射电子显微镜法	XRF	X射线荧光光谱法

3. 试验用器材、容器类英文名称

absorbent cotton　脱脂棉
agate mortar　玛瑙研钵
alcohol burner　酒精灯
asbestos-free wire gauze　石棉网
bath　冷、热浴
beaker　烧杯
blast alcohol burner　酒精喷灯
boiling stone　沸石
brown glass burette for alkali　棕色滴定管（碱）
brown glass burette with glass stopcock　棕色滴定管（酸）
cast-iron ring　圆形漏斗架
clamp regular holder　双顶丝
condenser　冷凝器
conical flask　锥形瓶
crucible　坩埚
crucible clamp/tong　坩埚钳
cupel　烤钵
distilling apparatus　蒸馏装置
distilling flask　蒸馏烧瓶
double-buret clamp　蝴蝶夹
dropper　试管架小滴管
dropping bottle　滴瓶
evaporating dish　蒸发皿
evaporator　蒸发器
extension clamp　万能夹
filter paper　滤纸
flash point tester　闪点仪
flat jaw pinchcock　止水夹
flint glass burette for alkali　白滴定管（碱）
flint glass burette with glass stopcock　白滴定管（酸）
flint glass solution bottle with stopper　白细口瓶
forcep　镊子
funnel　漏斗
glass rod　玻璃棒
graduated pipette　刻度移液管
heater　电炉
heating mantle　微波炉电热套
iodine flask　碘量瓶
iron support　铁架台
lab jack　升降台
lab spoon　药匙
litmus paper　石蕊试纸
magnetic stirrer　磁力搅拌器
matrass　卵形瓶
measuring cup　量杯
measuring flask/measuring cylinder　量筒

mercury-filled thermometer 水银温度计
oven 烘箱
pestle 研磨棒
pipetrack 移液管架
(one-mark) pipette 移液管
plastic wash bottle 洗瓶
porcelain 瓷器
power basic stirrer 电动搅拌器
reagent bottle 试剂瓶
respirator 口罩
respirator/gasmask 防毒面具
retort 曲颈瓶
rubber suction bulb 洗耳球
rubber tubing 橡胶管
scissor 剪刀
specific gravity bottle 比重瓶
stainless-steel beaker 不锈钢杯
still 蒸馏釜
stirring device 搅拌装置
stirring rod 搅拌棒
stopcock 玻璃活塞
stopper 塞子
stopper borer 打孔器
stopwatch 秒表
test tube 试管
tube rack/test tube holder 试管架
universal pH indicator paper pH 试纸
volumetric flask/measuring flask 容量瓶
watch glass 表面皿
weighing bottle 称量瓶
weighing paper 称量纸